"The nose on your face is the Buckingham Palace Guard of your body, the maître d' of all taste, as well as the seducer of your imagination and memory—and Jude Stewart has charmed them all into a wicked, poetic, and illuminating tour of their mysterious domains."

—Jack Hitt, author of *Bunch of Amateurs*

"In Jude Stewart's witty, surprising, and jubilant romp through the most undervalued of the senses, you will encounter some of most rapturous descriptions of scent ever put on paper. But even more than that, *Revelations in Air* is an expansive and large-hearted consideration of friendship, memory, beauty, and the art of appreciation."

—Adrienne Miller, author of *In the Land of Men*

PENGUIN BOOKS

REVELATIONS IN AIR

Jude Stewart has written about design and culture for *Slate*, the *Believer*, *The Atlantic*, *Fast Company*, *Design Observer* and other publications. She is also a contributing editor at *PRINT* magazine. She is the author of *ROY G. BIV* and *Patternalia*. Stewart lives in Chicago.

REVELATIONS IN AIR

A GUIDEBOOK TO SMELL

Jude Stewart

PENGUIN
BOOKS

PENGUIN BOOKS
An imprint of Penguin Random House LLC
penguinrandomhouse.com

LIBRARY OF CONGRESS CATALOGING-IN-PUBLICATION DATA
Names: Stewart, Jude, author.
Title: Revelations in air : a guidebook to smell / Jude Stewart.
Description: New York : Penguin, [2021] | Includes bibliographical references.
Identifiers: LCCN 2021007902 (print) | LCCN 2021007903 (ebook) |
ISBN 9780143135999 (hardcover) | ISBN 9780525507604 (ebook)
Subjects: LCSH: Smell. | Smell—Physiological aspects.
Classification: LCC QP458 .S74 2021 (print) | LCC QP458 (ebook) | DDC 612.8/6—dc23
LC record available at https://lccn.loc.gov/2021007902
LC ebook record available at https://lccn.loc.gov/2021007903

Printed in the United States of America
1   3   5   7   9   10   8   6   4   2

Book design and illustrations by Daniel Lagin

For Joy

One of the best little stinkers I know

CONTENTS

INTRODUCTION:
WHY A BOOK ON SMELL?

"My genius lies in my nostrils."

—*Ecce Homo* by Friedrich Nietzsche

S MELL IS A TESSERACT, COLLAPSING SPACE AND TIME. IT UNLOCKS MEMO-ries and grants us access to scenes we can enter only in imagination.

Can you revisit your childhood? Not actually. But walk into your old elementary school, and you'll be floored by how it smells exactly the same: dry chalk, wet wool, stale waft of cafeteria lunch. Sights and sounds will have changed beyond all reckoning, but smell is deep, transporting, and constant.

What does the inside of your body smell like? Surgeons say a healthy body's interior smells of almost nothing, just humid pulsing. It's the surgeon's knife that brings forth the smells. Blood emits a metallic tang; cutting through bone smells like burning hair.

What did World War I smell like? According to survivors, mustard gas smells of lilacs, garlic, horseradish, and onions. When the gas shell first hit, you'd smell nothing. Only as the gas evaporated would fresh springtime smells waft upward, like thoughts sharpened to lucidity: *Move!* The sharp mustardy note would materialize last, as a final warning.[1]

Outer space contains no air and therefore no smells. Yet space artifacts, returned to earth, smell eerily familiar. Lunar dust smells like spent gunpowder.[2] Mars dirt smells acrid and sulfuric, with a chalky, sweet undertone.[3] Phosphine gas found on Venus—suggesting life might exist there—would make the planet reek of decaying fish, assuming we could breathe on Venus at all.[4] If you could get close enough to smell them, Jupiter's inner layers would smell of bitter almonds, its outer layer of ammonia—the same smell that, perceived on a person's breath, hints at failing kidneys.[5]

Other diseases announce themselves by changes in bodily smells. Typhus smells of freshly baked brown bread, tuberculosis of stale beer, yellow fever of the butcher's shop, plague of over-ripe apples. Diagnosis by smell is both quaintly outmoded—how recognizable is measles today by its smell of freshly plucked feathers?—and relevant again. Trained doctors and dogs, and increasingly electronic "noses," can detect the smells of Parkinson's disease, melanomas, multiple sclerosis.[6] All around us, the air scintillates with smells, with information about change.

We smell with our entire bodies, and what we detect influences our lives more than we realize. Olfactory receptors line our noses but also our skin, skeletal muscles, and major organs. When exposed to a sandalwood odor, skin abrasions heal faster and hair regrows.[7] Clinical trials suggest we can detect one another's emotions by smell alone: joy, fear, disgust.[8] The sudden emergence of a phantom smell—perceiving an odor that's not actually present—can signal a developing brain tumor or the onset of an epileptic seizure or migraine.[9] Older people who abruptly stop detecting certain smells, like fish or peppermint, are more likely to die within five years.[10]

Because smell is only detectable at close range, it's intimate. Even when you think you're smelling something far away—like exploded fireworks or New York City hot dog carts—you're not bionically smelling across great distances. You're smelling a dense carpet-bomb of smell that's dispersed enough to reach your nose. And your worst fear about awful smells is true. The smell—that is, tiny airborne molecules from the smelly substance—really *does* invade your body's interior and bind to your olfactory receptors deep inside the nose. Smell's intimacy isn't a problem when you're sniffing something pleasant. But smell is largely involuntary: if you're breathing, you're smelling. Pungent smells can signal an unnerving proximity to food, to sex, to garbage, to pooping—to all the awkward, vital things.

That awkwardness is powerful. Smell is judgy by design; it patrols our body's boundaries for incoming threats. Smell infiltrates our oldest, reptilian-brain structures first, mashing all the buttons on the limbic dashboard: emotions, memories, opinions. But there's a built-in delay between smelling and thinking. Reason and words—the neocortex's business—attach last to smells and matter the least.

Smell's hasty judgments can be disastrous. Consider the medieval concept of *foeter Judaicus*, the supposed "stench of Jews" later appropriated by the Nazis.[11] Or the racist belief, not uncommon among whites until the 1970s, that Black Americans could take pills to reduce their bodily odor.[12] Any slight differences in bodily smells between ethnic groups have been exaggerated and weaponized by majority groups to justify racism and oppression.[13]

On the flip side, smelling can also bring people closer. In many cultures, choosing to smell a stranger signals a beautiful, mute friendliness. Many Arab cultures greet each other via "nose kisses": you press the bridge of your nose against the other person's, then inhale each other's scent.[14] New Guineans insert a hand into the other person's armpit, then sniff their fingers appreciatively.[15] An ancient Indian text echoes a greeting from that part of the world: "I will smell thee on the head; that is the greatest sign of tender love."[16] Because it requires vulnerability, smell engenders trust.

Smell is intelligent, too. The number of unique smells humans can distinguish may go as high as one trillion. We aren't the best all-around smellers in the animal kingdom, but we are the best discriminators between smells—thanks to our brains, huge portions of which are wired for smell. Why devote so much brainpower to smell? Hard to say, but smell-psychologist Rachel Herz offers an intriguing theory: "The ability to experience and express emotion grew directly out of our brain's ability to process smell." Herz means this literally—the human brain's amygdala slowly evolved out of the primitive olfactory cortex which controls smelling. But she's also referring to evolution of a metaphorical sort: "Emotions are to us what scents are to our animal cousins," she writes in her book *The Scent of Desire*. "Smell for animals informs survival in direct and explicit ways; for us its primary survival codes have been transformed, into our experience of emotions." Herz calls this phenomenon "olfactory-emotion translation."[17]

Smelling takes a lot of work, biologically speaking. The process of smelling is complicated and only approximately understood. Humans have more than four hundred olfactory receptors, compared to only four visual receptors. While scientists discovered the auditory and visual equivalents centuries ago, olfactory receptors were just discovered in 1991. How our receptors detect smells is a particularly deep riddle. A molecule's shape seems to determine which smell receptors it binds to; beyond that we have no earthly idea why molecules smell as they do.[18] Take the molecule benzaldehyde. Add a double-bond of oxygen to its tail: the molecule becomes cinnamaldehyde and its smell shifts from almonds to cinnamon. Now lengthen cinnamaldehyde's tail with another five-carbon atom chain, and the smell shifts to floral. There's no pattern scientists can discern—yet.[19]

Put another way, you could say smell lacks an alphabet. Unlike the wavelengths organizing the audio or visual color spectrums, smells share no known common denominator.[20] Thus the landscape of all smells is enormous and resists easy reduction.

This complexity explains why we won't transmit smells via the internet anytime soon. Smells *can* be passively detected: electronic noses monitor food factories for spoilage and nuclear reactors for disastrous leaks.[21] But two-way transmission of smells is still impossible. Valiant (if gimmicky) attempts to transmit smells—like Cyrano, the 2016 "digital smell speaker"—are basically cheats.[22] What's transmitted isn't a dynamically remixed smell, just an electronic signal to release a prepared vial of scent on the receiving end. Smells can be digitized and recorded, but they can't be mediated by tele-communications. You can only smell things live.

Think about that: Smell delivers us back to our bodies, underscoring liveness. Isn't that the same as underscoring life itself?

LIVENESS IS WHAT DREW ME TO SMELL, and to writing this book. In 2016, on our last night of a summer spent in Berlin, our friends Heidi and Thomas enthused about visiting an art exhibit called Smeller 2.0. A sealed room equipped with a hidden contraption resembling a church organ, Smeller 2.0 delivered a stream of smells in quick succession. These could be synchronized to movie scenes or live performances in the room or emitted as standalone compositions. Smeller 2.0 claimed to improve on previous smell technologies in two ways: by offering a wider range of smells formulated with a perfumer's help; and by better controlling ventilation in and out of the space (so smells wouldn't linger or inadvertently mix).[23] The idea stuck in my mind.

Reading about Smeller 2.0's historical antecedents hooked me further. The dream of layering smells onto movie-watching is nearly as old as movie-watching itself. This history is a pileup of interesting failures, technical experiments, and overweening optimism in smell-o-vision's eventual appeal. Despite numerous iterations of smell-tech, audiences kept complaining: the smells were fake, unconvincing, too strong, too weak. The smells were poorly synced to the movie's scenes. After several showings, theaters accumulated a strange, off-putting amalgam of scents. Most interestingly, people who'd bought tickets to see these movies seemed sneakily pleased when smell-o-vision failed.[24]

Why smell-technologists kept pursuing this dream seemed intuitive: smell hits you where you live. What's more immersive than smell? But that fact also explained our defensiveness. Unlike vision or hearing, smelling is a private sense. Whether a shared movie experience or tantalizing scents pumped into a food court, our resistance to manipulation by smell is mute but real. We *will* fight the smell-proffering robot overlords.

My response to big, inarticulate questions is to try to write about them, but I had never really thought much about smell before—which is curious for an everyday phenomenon capable of triggering the most visceral of emotions. And so I researched, digging into newer smell-transmission technologies, from VR to iPhone contraptions—all still unmatched against our weirdly discerning sense of smell, both capable of detecting false notes with exquisite accuracy, yet incapable of describing smells in words[25] or remembering them without direct stimulus. If our genius lies in our nostrils, as Nietzsche remarked, it's an untrained, lopsided genius at best. Smell is a latent superpower I'd previously ignored, not just right under my nose, but *inside* it.

My untrained sense of smell couldn't contrast more with my highly trained sense of vision. As a design writer, I've done plenty of professional seeing and really enjoyed it. But lately the fun of visuals has steeply diminished. Images have become too perfect, nutrient-lite, abundant and cheap. We're constantly both making and consuming them. They've become accelerants, hurtling us toward shopping. Consuming each picture takes zero time and yet collectively devours hours. We like images now not necessarily for themselves, but because looking at images precludes thinking. Taking a photograph seems to answer, falsely, an existential question.

With this surfeit of images comes anesthetizing, a sneakily attractive part of the visual trap. Visual thinking is cerebral, monetized, and ultimately defensive. It pushes the embodied world away, allows us to retreat into our brain-jars, tiny flickering screens propped up near them. Tap, tap. Blink, blink. Seeing is soothing, yet irritating even in rest.

Prior to Smeller 2.0, my attention—my literal and metaphorical eye—had already been floating away from images, and more toward whatever the deepening blizzard of images was papering over. What other sense perceptions do images help me hide from? Don't get me wrong: I still liked graphic design, visual art, looking and seeing. But I wanted *more* from my senses.

And so I began to wonder: How could I become a better smeller—and how might my life change if I did that? If a smell can encapsulate a personal memory, can it also encapsulate a collective history? And why does an avid interest in smell seem a little embarrassing?

Along with its liveness, smell's awkwardness drew me like a magnet, a fight of both pull and push. Imagine walking around the world, sniffing everything you encounter like a nose-forward handshake. Weird, right? The awkwardness of approaching life in smell-forward terms really clinched it for me. *Follow the awkwardness* sums up my philosophy as a writer pretty handily. To me, finding the words to unlock strongly felt sensations in another register—whether color, pattern, or smell—thrums with life and potential.

I'm not alone in this way of thinking. In a 2018 *New York Times* op-ed titled "We Have Reached Peak Screen. Now Revolution Is in the Air," Farhad Manjoo declared, "tech has captured pretty much all our visual capacity." He calls for a "less insistently visual tech world . . . to take some pressure off our eyes."[26] Audio's rebirth does just that. Music's popularity in live venues stems from our desire to feel the beat inside our bodies, to make the auditory haptic. (Changing economics matter, too: we won't pay for music itself anymore, but we will pay to experience it live.) Podcasts fill our ears with ambient noise, with the urgency of unplanned speech. We relax into listening. Audio takes its time and builds.

Taste is resurgent, too. What sense could have been more humdrum before foodies started telling the many stories manifest in foods? We're rediscovering touch, too. Think of sinking your hands gratefully into bread dough after a lightning round of coding, or reclaiming analogue skills like knitting or carpentry. Making tangible things unlocks a nonvirtual way of being.

Smell deserves its turn, and I'd argue its cultural moment is here. Overlapping with taste yet larger in scope, smell is the sense that comes closest to pure perception. Smelling demands concentration and openness, qualities many people are keen to recultivate. The American science writer Lewis Thomas once observed: "The act of smelling something, anything, is remarkably like the act of thinking itself."[27]

Yet smells don't happen purely inside the mind. Smell brings us into direct, uncanny contact with the world. "If everything were smoke," wrote Heraclitus, "all perception would be smell." In a funny way, that's an accurate description of what's happening. Smell is what results when any substance transmogrifies into air. Spores of the world stream nonstop past our noses. It's an ever-changing flux, recalling Heraclitus's other famous remark about time: "You cannot step into the same river twice." To enter the great stream of smells—a stream connecting present with past, people and far-flung cultures with each other, all urgently and wordlessly—all we need to do is sniff.

Why write a book about smell—or read one? To challenge myself and readers to use a sense that's barely understood. To find pleasure and unfamiliarity inside my own body again. To feel less expert, more vulnerable, slower, live, and awkward. I hope *Revelations in Air* will satisfy these desires while also stoking them for you. I want it to entrain your curiosity in an ever-widening arc, plumb the strange outer limits of smell, and take seriously what many consider a deeply unserious subject. I want to lean into the corniness of smell, too. Noses are ridiculous, oddly genital, parked in the middle of our faces. Sniffing is childish, feral, rude, inappropriately sexy—but startlingly beautiful in flashes. Smell is perfectly gonzo.

HOW TO READ THIS BOOK

VERY GUIDEBOOK STARTS WITH A PRIMER ON EQUIPMENT AND PROPER methods of using it. Just like birdwatchers use binoculars and waterproof notebooks, your equipment for smelling is your nose. Part 1, "The Nose," tells a concise history of the smeller itself: what this equipment can do and how we've used it historically. It acquaints you with your nose's lesser-known features and trains you to use it more expertly.

Properly equipped and ready to sniff, next we take our noses out into the wide field of smells. Inside the scent of a tea-soaked madeleine, Proust found his entire childhood folded, like an impossibly detailed origami. In a similar spirit, I wanted to find out how many stories an individual smell could contain, how deep each smell might go. But which smells should I choose?

Because smells lack an alphabet (or at least, one we can decode yet), smells resist categorizing. The best approximation, categories from the perfume world, isn't ideal since that system exists mostly to organize lovely smells. I wanted to explore *all* the smells—or at least a wider gamut, stinks included. Ultimately I adapted scent categories recently established by smell-researchers at Bates College and the University of Pittsburgh.[1] My categories include flowery and herbal; sweet; savory; earthy; resinous;

funky; sharp and pungent; salty and nutty; tingling and fresh. I also included a tenth category, otherworldly, for smells that defy easy categorization.

Each chapter tells a detailed story of one smell, but also unlocks a broader aspect of this mysterious, lower-order sense. First you'll read a close description of the smell in that chapter. Before you go too far, I encourage you to find that smell yourself and smell it live as you read. Smelling is a kind of meditation turned inside out. Its focus is not inward on breathing, but outward on your surroundings. Breathe the smell in and hold a moment of wordless concentration. You'll enter a different sense register. Then plunge on to read more about it.

Revelations in Air is punctuated with one-page exercises to practice your smelling skills at home and ends with an FAQ. What are the most universally loved—and reviled—smells? Has anyone copyrighted a smell? Do other languages label smells that English speakers can't name directly? This book can tell you.

Before I wrote *Revelations in Air*, I thought air was empty; now I know otherwise. Air churns with invisible action, revealing smells with all the extraordinary information they impart—*if* you're paying the right kind of attention. Here's hoping you enjoy tuning into smells and find many revelations in the air that surrounds you.

REVELATIONS IN AIR

Part One

THE NOSE

1. MEET YOUR NOSE (AND MINE)

What is the nose? Most obviously, it's the bony edifice parked in the middle of your face. It's a pedestal for sunglasses, an awning for your mouth that you smear on sunny days with zinc oxide. If you choose, you can pierce its nostril with a jewel or the cartilage of your septum with a bull ring. At its arty-cinematic best, a nose might become a promontory glazed by a single, crystalline tear. Surgically, noses are candidates for downsizing, never for upsizing. Seen in profile, the nose sails forth in triangular glory, a prow anxiously hoping to be balanced out by a chin or undercut by the backward sweep of hair. Seen head-on, the nose disappears again, a 3D shimmer midface edged with two dark holes.

Most of our thinking about noses focuses on their outsides, not on their vast interiors and what our noses can or cannot do. At least, that's how I've always considered my nose—when I've considered it at all.

My own nose is Roman, single-humped but straight, not pierced, and definitely not small. When I tilt my face upward and look at the sky, you can see my nose's most remarkable feature: its point isn't a point at all, but square-shaped and flat. Seen from the underside, my nostrils resemble a two-pronged electrical socket for some country wired more crudely than my own.

I broke my nose once, a clumsy moment in an otherwise romantic story. In college during winter break, I slogged through a heavy snowstorm to catch a train that would deliver me to see Seth, the man I later married. I reached 30th Street Station in Philadelphia with only minutes to spare. Beelining it to the train track, I burst through the swinging gate doors with such force, the doors smacked me back right on the nose. It hurt with a deep pain that irradiated my face thoroughly for five minutes. Soon a faint purple horizontal line bisected my nose's thumping, raised bridge. The line eventually faded but my nose's new elevation did not.

I like my nose now. In choosing not to get it "fixed," I'm trading a good-enough nose for an even better story. My new, larger nose scrambles my ethnicity nicely, makes me register less clearly as an American ex-Catholic of German extraction. My nose looks broken-in because it is. Before I broke my nose, I was already a tall, solidly built woman who knew no amount of slouching would make me seem petite and super-feminine. I've learned to become okay with taking up more physical space than women usually do. And just like a larger-framed body, a sonorous voice, or a bigger personality (and I have all of these), I cannot hide my sizable schnoz. It demands acknowledgment and acceptance.

Before I began work on this book, I had no idea how good of a smeller I was. Let's say I aspired to normal capabilities and felt myself within range. I can usually perceive smells other people remark on, and I'm sometimes the first person in a group to notice a smell. I'm enough of a foodie to know I'm not a slouch in the smeller department. But, having easily irritated skin, I'd never gotten into perfumes. Incense, scented candles, aromatherapy—I'd always felt studiously neutral about these things. I'd had only brief flashes of super-smelling while pregnant. It happened only once, in fact: I recall smelling a sizzling flank steak, in stereoscopic detail, across a restaurant crowded with other meals and smells. So I came to this book as decidedly inexpert but eager to train myself.

Pictures of disembodied noses bumble gracelessly across so many books about smell, from their covers to their interior pages. Yet nobody talks about how funny that fact is, or why it also makes us feel uncomfortable. Detached from its face, a nose always seems like it should have two dangling strings attached. It's a mask of personhood, even a synecdoche for the entire self.

Considering the nose independently as universal yet nonstandard equipment raises questions. Is the nose ennobled or absurd? Two air-holes connecting the head to the lungs, or the soul's spiritual grate? Is the nose one of the most animal things about us, or the most human? Should we laugh at its powers, channel them, even fear them? How deep do those powers run?

A book about smell should start with noses for more than the obvious reasons. Noses are funny and awkward and fascinating. They're also surprisingly individual in capabilities and predilections. In their symbolism across cultures and history, noses make concrete a lot of mute, contradictory feelings that swirl around smell itself.

2. FIDDLE WITH THE EQUIPMENT, OR: HOW OLFACTION WORKS

Your equipment for smelling is your nose. Yet most of us wield this equipment without knowing much about it.

When we smell something, what are we actually doing? Like taste, smelling—or olfaction—is a *chemical* sense. When you smell, you're sensing airborne chemicals, or volatile molecules, wafting through the air and up your nose. Those chemicals carry rich and highly specific information pertaining to matters from survival to pleasure. Smells can tell us about encroaching threats, like a wildfire or whether foods are delicious (or dangerously "off"); they can also help us find alluring potential mates. (For more on the smell of other humans, see the skin section in chapter 6.) Even when you're unaware of smelling something, a scent can unconsciously confirm and amplify the evidence of your other senses, punching up the detail and realness of perception.

Not everything gives off a scent. Why? A substance emits smells only if stray molecules are prone to get knocked off and take flight. This is called "becoming volatile" and usually happens to molecules during chemical reactions—for instance, when they're heated—or as a result of microbial metabolism, which releases odors into the air. Glass, steel, porcelain, most rocks—their molecules aren't prone to volatility and thus don't usually smell like much.

So how does a smell reach us? To float successfully to our noses, a smell molecule—or "odorant"—must be buoyant and weigh less than three hundred Daltons. The weight of one atom of hydrogen is referred to as a Dalton, after the English physicist John Dalton. So an odorant molecule must weigh less than the equivalent of three hundred atoms of hydrogen for us to smell it. Smells are buoyed aloft by turbulent airflow but smell molecules also obey gravity. Absent any airflow, odors will accumulate near the ground and collect around objects. Smells may be invisible, but that doesn't make them any less physical.[1]

Once the smell reaches your nose, the equipment really kicks into action. You inhale, and odorants enter the nose. This might happen passively or actively. Your sniff pattern—to help you actively perceive a smell—is highly individualized and distinctive to you. Midway through the first sniff, your nose has already zipped enough information about the smell to the cortex to optimize your subsequent sniffs and capture the smell most efficiently. Thanks to your sniff pattern, you'll rarely overwhelm yourself with sniffing too much stink or exhaust yourself in smelling something lovely.[2]

The inhaled air gets warmed, filtered, and humidified as it bumps its way through the turbinate passages in your nasal cavity. This smell-laden air eventually reaches a spot a few centimeters behind the area where eyeglasses rest on your face. Here is where the olfactory organs, stacked directly under your brain's frontal cortex, reside. At the bottom of this stack is your olfactory epithelium, a yellow, mucous-rich layer.

Inside the sticky, moist epithelium nestle the olfactory neurons. They dangle downward, like upside-down carrots growing in soil. Smell molecules first stick to the epithelium and then melt, perhaps with a sigh of arrival. From the millions of smell molecules in the outer atmosphere, our noses siphon up only a few hundred or thousand smell molecules with every sniff that eventually reach this final destination. The olfactory neurons detect the smells, but it's the olfactory receptor proteins sitting inside the neurons that actually bind to odorants.

As mentioned previously, how an olfactory receptor detects a specific smell is a deep riddle. The smell-molecule's shape seems to determine which olfactory receptors it will bind to; most olfactory scientists liken this to a key fitting a lock.[3] But each olfactory receptor can detect many, many different smells to varying extents because they're promiscuous binders to smell molecules. You could think of each olfactory receptor as an extremely loose lock. Many, many keys (or smells) will fit into a given receptor and open the door successfully, some requiring more key-wiggling than others.

This is clearly oversimplifying matters because we've got about six million receptors across four hundred different receptor types, yet our noses can distinguish between many more smells than that. How many? Estimates range from eighty million to a theoretical upper limit of one trillion smells.[4]

Moreover, those four hundred olfactory receptor types aren't a uniform, standard-issue set. About a third of my receptors will differ from yours, and each of us may have specific anosmias—smells we can't detect—that we may or may not be aware of. Yet most of the time two people will inhale the same lily's scent and identify it accurately, irrationally confident that they're both experiencing the smell the same way.[5]

However it happens, the smell molecule finds its matches among the olfactory receptors. These fire an electrical signal to the olfactory bulbs, two buds that hang from a tongue-like protrusion of nerves connecting to the limbic center of the brain. This chain of nerves rests atop the olfactory epithelium and is protected from the outside world only by a thin layer of mucus. Olfactory neurons regenerate every four to eight weeks and change over time, responding to whatever smells they encounter most often.[6]

The olfactory bulbs are thought to be the brain's primary processing center for smell. The bulbs take in information from the olfactory receptors, encode it into a unique odor signal, and then pass this signal to the olfactory center in the brain's cortex. I say "thought to be" because this basic relay of signals has recently been thrown into doubt by a highly controversial study. While conducting MRI brain scans at the Weitzmann Institute of Science in Israel, researchers scanned a woman who demonstrated a normal sense of smell despite the fact that she completely lacked olfactory bulbs. Since the subject was both female and left-handed—two traits that sometimes influence how the brain is organized—the researchers invited other left-handed women to get their brains scanned. Their initial results suggest up to 4 percent of left-handed women may lack olfactory bulbs but can still somehow detect smells normally.[7]

Smell gets even weirder than that. Remember why olfactory receptors exist in the first place: to help our bodies sense chemical fluctuations important to our health or safety. The genes that encode smell receptors also happen to be expressed by many other body parts. Kidneys can "smell" signals from your gut bacteria and moderate your blood pressure after a particularly heavy meal. Swimming blindly in a silent void, sperm are guided towards the egg by the latter's alluring smell. Because olfactory receptor cells can regenerate, scientists are attempting to use them to heal spinal cord injuries. Your lungs, blood vessels, muscles: growing evidence suggests they are all constantly smelling.[8]

Our scientific understanding of olfaction is still pretty cursory, and there's a lot more to know. As we move from the nose deeper into the brain, it'll become clearer why smells are so deeply entangled with memory and emotions.

3. SMELL AS EMOTIONAL TIME TRAVEL

Why do we learn to smell in the first place? How early in life do we start?

We meet our first smells in the womb at twelve weeks of gestation.[9] Our sense of smell patrols the body's most vital boundaries and pushes our protective force field beyond our skin. The olfactory system is trying to detect any chemical that could hurt us before that chemical reaches us at full strength. It's a tall order.

So the fetus practices, by breathing in amniotic fluid and smelling its contents: not only what the pregnant mother eats and drinks, but the creams and shampoos she uses, even smells the pregnant mother inhales herself. (Smells travel through liquids much as they travel through air.) From pre-birth onward, we encounter fresh smells constantly, gleaning information about each smell and eventually making judgments about them.[10]

Those judgments are swift, binary, and uncompromising, as one study with rats demonstrates.[11] First, researchers taught female rats to fear the smell of peppermint, then got those rats pregnant. The mama rats gave birth to babies, who were then exposed to the smell of peppermint *and* the smell their mother gave off in the presence of peppermint. In effect, researchers were pairing the peppermint smell with mama-fear smell in the babies' minds. Nearly all the baby rats learned to fear peppermint's scent after smelling it just once. And their aversion was extreme: many babies tried to block the tubing that emitted the awful minty odor. Humans suffering from osmophobia— fears triggered by certain smells—have similarly paired a smell with an early childhood trauma.

Human brain circuitry explains why smell, emotion, and memory are so tangled. For most senses, a stimulus travels from the sensory organ to the brain's thalamus. The thalamus is considered "new brain": more evolutionarily recent, new-brain regions are responsible for complex, higher-order processing skills.

Smelling is different. Smells bypass the thalamus with all its fancy new-brain abilities like language and logic. Instead smells move directly from the nose to the olfactory bulbs, where smells are processed in the brain. The olfactory bulbs are enmeshed with the amygdala and hippocampus, parts of the primitive "old brain." The hippocampus handles memory, specifically *episodic* memories that constitute our life's personal narrative. The amygdala handles emotions: registering them, regulating them, fulminating over them.

Important memories in your life are nearly always emotional. When an important episodic memory forms, we're feeling all the feels, registering all the details clearly. If we happen to smell something distinctive during that experience, the amygdala, hippocampus, and olfactory bulbs fuse feelings, memory, and smell together effortlessly.[12]

The other senses are processed, and subdued, by the new brain. Smell is almost completely old brain: its sensations are rawer, unverbalized, purposefully indelible. Smell and old-brain are so interconnected, in fact, that scientists often refer to the entire primitive brain as the *rhinencephalon*, or "nose-brain."

Study after study confirms our hunches: odor-induced memories date from earlier in life and conjure a stronger emotional response than memories evoked by other sense stimuli.[13] Smell is so reliable, actors often sniff something evocative backstage to trigger the right emotions for a performance. (For more on smells and memory, see the pencils section in chapter 5.)

In short, your nose is both a skeptic and a pessimist: it judges new smells swiftly and absolutely. Unfamiliar smells are presumed bad unless explicitly proven otherwise. Your brain is constantly filing away these judgments, an ever-growing smell database for future reference.

4. HOW MANY SENSES ARE THERE?

We're used to thinking of smell as one of five classic senses, the other four being sight, hearing, taste, and touch. But in fact humans have way more than five senses—the current scientific count ranges from fourteen to twenty senses. Some of these senses are finer gradations of the five senses we already know. For instance, the broad category of touch includes three related senses: thermoception (ability to sense heat or cold, even without directly touching something), a sense of pain, and a sense of pressure.

Other senses don't map easily onto the "big five" at all. Take proprioception, or our awareness of where our body parts are in space. (Close your eyes and touch your finger to your forehead. You can do this effortlessly thanks to proprioception.) Equilibrioception helps us maintain our sense of bodily balance. Kinesthesia is our awareness of motion, chromoception our sense of time passing. (Ever wondered why you can guess what time it is with uncanny accuracy? That's your chromoception at work.) Other species have additional senses that elude us humans. Sharks use electroception, an awareness of electrical fields, to hunt down prey. Homing pigeons and bats can navigate based on magnetoreception, or the ability to detect magnetic fields.[14] Babies, like bats, can orient themselves via sonar, making sounds and then observing how those sounds bounce back to them.[15]

But really no sense operates in isolation from the others. Put another way, our senses are always amplifying each other's perceptions, operating in sync to build a detail-rich picture of the world and our experiences within it. It's difficult to balance on one leg (i.e., use your equilibrioception) while blindfolded (i.e., without vision). Similarly, you can readily differentiate between a photo of a steaming-hot cup of cocoa and an actual cup by smelling the aroma and grasping the warm cup in your hands. Some of these sense-interactions are downright odd. For instance, one group of scientists recently found neural evidence in mice of "smounds," smells whose perception is directly affected by hearing a particular tone at the same time.[16]

Smelling often blurs with taste, another chemical sense. As you chew, food smells waft inside the mouth directly into the nasal cavities via the throat—what's known as retro-nasal smelling. That concentrated, detailed hit of smell is what actually makes food taste good and keeps you eating.

Our sense of taste is surprisingly crude. Your tongue can perceive five tastes—salty, sweet, sour, bitter, and *umami*, a Japanese term describing a fatty or savory flavor—but that's it. Hold your nose and then sip some coffee, and you'll only get a bitter taste. But release your nose to the smells *while* tasting, and then you'll identify this bitter taste specifically as coffee—or perhaps even more specifically as latte macchiato or espresso. We depend on our sense of smell to render broad taste classifications higher-res and pinpoint exactly what we're enjoying. *Flavor* is the overarching term to describe how smell and taste merge with other sensory cues like mouthfeel, temperature, and even pain receptors (for spicy foods).[17]

Smell also has a physical dimension akin to touch. Smelling activates the trigeminal system of nerves in the face and nose; these respond to temperature, touch, and pain. All smells trigger some degree of trigeminal response, but it's usually faint. Smells known for a strong trigeminal response include cooling camphor (see chapter 5) and eucalyptus, blinding skunk spray (see chapter 7), stinging habanero peppers, burning ammonia, tear-inducing chopped onion. Often when you dislike a particular smell or taste, it's really the smell's trigeminal response that's off-putting.[18]

Where did the idea of five physical senses originally come from? And how do we explain the hierarchical pecking order that's evolved between them? These are two huge philosophical questions that we'll answer only briefly below. It was Aristotle who identified five physical senses to correspond to the five natural elements—earth, air, water, fire, and "quintessence," a sort of indefinable soul matter. Each of Aristotle's five senses (except quintessence) also neatly corresponded to a conspicuous sense organ in the body. Aristotle ranked vision at the top of his sense-hierarchy but accorded smell an interesting unifying role; he considered that smell linked sight and hearing with taste and touch.[19]

Aristotle's framework has stuck around the Western world for a long time, but its durability obscures centuries' worth of debate about the exact nature and number of human senses, how trustworthy is the evidence of one's senses, and so forth. Philosophical arguments have erupted over whether the power of speech or the emotions should count as senses. Medieval Christian theologians discussed five inward senses, spiritual cousins to the physical senses. These included memory, instinct, imagination, fantasy, and "common sense" (a mental faculty that processed information gathered by the five physical senses). In a delightful seventeenth-century allegorical play *Lingua*, the figure of speech (Lingua) wishes to become one of the five physical senses. Common Sense presides over a court case in which Lingua's claim to sensehood will be decided. Praising each sense in turn, Common Sense declares Smell "the high priest of the Microcosm" before rendering a verdict: Lingua cannot be an official sense because they cannot number more than five. That is, Aristotle wins again.

In general vision has always been top dog among the senses, but smell has had its stalwart defenders. These include Aristotle's predecessor Diogenes; the Roman physician Galen (who considered smell superior because it was the sense most directly linked to the brain); as well as Enlightenment-era philosophers John Locke, Jean-Jacques Rousseau, and Denis Diderot. Defenders of smell's standing in the sense-hierarchy cited its immediacy (that it revealed the literal essence of an object, in the form of airborne particles), its emotional force, and—by Christian theologians at least—its undeniable moral quality of distinguishing good from evil. But smell's unmediated quality also made it unusually potent among the senses and therefore suspect.

Smell's prestige declined in the West with the invention of the printing press, when oral culture gave way to a more visual, text-oriented culture. The Enlightenment period demoted smell still further. Thanks to philosophers like Immanuel Kant, Georg Hegel, and Friedrich Nietzsche among others, vision was deemed the most accurate sense, ideally suited to scientific pursuits and knowledge-gathering. How could smell— emotionally triggering, unstably present, difficult to describe or quantify—possibly advance the causes of science or industry the way vision could? Couple these concerns with the colonist backdrop of the day, and it's easy to see why Western thinkers were happy to denigrate smell to the most atavistic sense: antithetical to reason, keenest among children, women, primitive peoples, and animals.[20]

Fast-forward to the deodorized present, in which smells are engineered to remain unobtrusive. By one estimate, only 20 percent of the $30 billion–plus flavors and fragrances industry consists of luxury perfumes. The other 80 percent consists of "functional perfumes": the smells added to laundry detergents, cleaning products, and an infinity of household objects after removing any unpleasant natural or industrial smells. "Unscented" products are a stage-managed fiction: we take out any pungent or overtly bad smells and replace them with fainter, more pleasing ones.[21] Smell's standing among our five (or more) senses is still quite modest in the modern world.

And yet. However humble, smell's subversive powers are considerable. Historical attitudes toward the nose—the visible organ of smell—reflect a backhanded kind of reverence to our powers of smell and how this lowly sense might embody some of the finest qualities of our personhood: intelligence, personality, and even character.

The capacity of taste and smell have long been considered proxies for overall intelligence. The Latin *sapere* means both "to taste" and "to know"; thus *homo sapiens* is both "knowing man" and "tasting man." Similarly, the word *sagacious* stems from the Latin word *sagacis*, meaning "keen-scented," perceptive in a way that amasses wisdom. Then as now, we describe a wise person's assessments in metaphors of smell. They "smell" trouble or opportunity; they catch a "whiff" of the same; or they subject doubtful claims to the "smell test." The word *nose-wise*, now obsolete in English, conveyed both keen smelling abilities and cleverness.[22]

The nose wasn't only considered a seat of intelligence; it was also the organ where the soul resides. Classical Latin writers referred to a life-or-death moment by declaring "My soul was in my nose," i.e., on the verge of escaping the body permanently.[23] Across many cultures and historical contexts, smells have given humans a palpable means of communicating with the spirits and the dead, who supposedly dwell in the ether or the skies above. Burning incenses to please or even feed the gods makes smell a tangible vehicle of spirituality, with the nose as its witness. (For more, see the sections on camphor, chapter 5; cinnamon, chapter 2; oud, chapter 5; frankincense and myrrh, chapter 5; and old books, chapter 10.)

Given all this weighty symbolism in the middle of our faces, it's perhaps inevitable that people would try to link nose shapes to different personality types. Various nose-classification systems have attempted to systematize this "knowledge" much as the quasi-scientific theory of phrenology linked head shape and size to personality types. Many such nose-classification systems provided a quasi-scientific cover for racism; others were less malign but make for excellent satire.

The eighteenth-century experimental novel *Tristram Shandy* is huge on nose-classification discussion. Tristram's father, Walter, was obsessed with noses, collecting every treatise he could find about them. Believing one's nose announces a man's

Observe smells properly.

Crush materials in your fingers to see what smells you can release from it. You're coaxing a solid substance into becoming airborne, something that can travel inside your nose. Pulverize the material as finely as you can.

Water unlocks smells, so wet the things you sniff whenever feasible. Intrepid smell-hunters will carry around a spritzer bottle, but a light dab of water will also do. You'll particularly want to tune into smells when it rains or the humidity rises.

Cold deadens our ability to smell, so use your breath to warm up smells wherever possible. Hold whatever you want to smell in the palm of one hand. Cup the other hand tightly around it, then sniff through the gap in your fingers.

Waggle your nose and, if possible, waggle the smell itself. Perfumers and other professional smellers do both: waving a paper strip impregnated with smell under their nostrils and moving their own noses rhythmically above the smell stick. Smells splay out in tendrils, much like ink disperses in water. Waving the smell releases more odor molecules. Moving your nose while sniffing helps you find more scent-tendrils and observe the smell from all sides.

Pay attention to whether a smell feels pungent, chilly, or eye-watering. (Think of eucalyptus, skunk, chopped onions, or ammonia.) These smells are triggering your trigeminal nerve, the cranial nerve controlling facial muscles.

Bring a small, wide-mouthed glass jar to place your smells in. (Glass is best: it's impermeable and won't contribute any smell of its own to the mix.) You can insert a bit of moistened paper towel to amplify the scent.

greatness, Walter felt gutted when Tristram's nose was crushed at birth by forceps wielded by Dr. Slop.[24] Squashing Tristram's baby nose renders him hapless, intellectually unserious, and sexually unattractive to boot. Another character in the novel, Hafen Slawkenbergius, is a self-professed authority on noses (with an exceedingly long nose himself). Slawkenbergius has translated the "Neo-Latin" monograph *De Nasis* (*On Noses*), an imaginary treatise about how nose shapes reveal human character. In the novel Slawkenbergius is styled as too smart or "nose-wise" for his own good, with a name that's itself a stink-joke: his first name, *Hafen* means "chamber pot" in German, and his last name alludes to *Schackenberg*, or "manure heap."

A famous real nose-classification book was *Nasology*, published in 1848 by one George Jabet, probably a pseudonym for the author Eden Warwick. Also drawing from phrenology, *Nasology* claims that a person's character and intelligence can be intuited from their nose shape. Jabet identifies six nose types: three have largely positive qualities—Roman, Greek, and wide-nostriled or "cogitative." The fourth type, which Jabet terms the Jewish, commercial, or hawk nose, indicates worldly shrewdness and a knack for profitable insights. While Jabet names this nose-type after the Jews, he notes that it's not exclusive to them—anyone good at business may sport this nose. Jabet's last two types, the snub nose and the "celestial" nose (so named because it points upward to the heavens), indicate moral weakness. It's not accidental that an English writer of this period would judge the nose-type characteristic among the Irish as degenerate.[25]

Nose-classification systems have Eastern counterparts—for instance, in the Chinese tradition of nose-reading. A kind of bodily horoscope, nose shapes can supposedly signal many aspects of personality, especially one's potential wealth. Particularly during a person's forties, the shape of their nose can reveal how rapidly they can earn money and how well they will handle or invest it.[26] According to this system, my own large Roman nose means I prefer being my own boss, am impatient and high-energy, and handle money well. Check, check, check.

In these systems big noses often signal a sensual nature, and noses themselves are undeniably phallic. Sigmund Freud thought the nose and genitalia were linked and that sexual repressions might be cured with nose surgeries and cocaine-sniffing. In a very weird partnership with surgeon Wilhelm Fliess, many of Freud's patients underwent nasal surgery as a complement to psychoanalytic treatment.[27] Whiffs of Freud's thinking—that the nose is basically an unclothed penis—linger in literary references to big-nosed sensualist characters and sexuality's many delicious stinks. (For more on this, see the skin section, chapter 6.)

> If you are ambitious to found a new science, measure a smell.
>
> —Alexander Graham Bell

By extreme contrast, amputating someone's nose is a profound humiliation, an X-ing out of the person. Removing a person's nose as punishment for crimes is surprisingly prevalent in history: ancient Egyptians, pre-Columbian Americans, and early Indian and Arab cultures all practiced it. Rhinotomy (nose removal) was wielded against one's political opponents and most often punished abuses of power or unfaithfulness to one's spouse or superiors. The Roman emperor Justinian lost his nose this way when he lost the throne, but later wrested power back while wearing a golden nose prosthesis.[28] The sixteenth-century astronomer Tyco Brahe lost his nose in an ill-advised duel and also later wore a gold prosthesis. So many noses were chopped off as a mass punishment for insurrection that the Nepalese city of Kirhipu earned the nickname "The City of the Chopped-Off Noses." A similar nineteenth-century event lost Turkish soldiers their noses to Bulgarian invaders, but luckily the victors distributed silver prostheses widely when amnesty was declared. Unsurprisingly, historical advances in rhinoplasty—the surgery we call a nose job—paralleled these waves of amputations.[29]

Perhaps more than any other body part, your nose represents you as a fully-faceted person: intelligent, spiritual, idiosyncratic, and sovereign. For perfumers, this equivalence is also professional: perfumers are nicknamed "noses," as in, "Jane Roe works as a nose for Firmenich." Professional noses undergo rigorous training to identify successively finer gradations between perfume ingredients, learning which molecules are responsible for which smell-notes, assimilating common smell combinations (called "accords," in a nod to musical chords), and then putting all this knowledge to work to create their own fragrances. Noses work closely with fragrance chemists—for more, see the section on invented smell, Iso E Super(chapter 1). The sections on rose and jasmine (chapter 1) explore the history and technology of perfume creation, too.

6. SMELLING THE FUTURE

While research into human and animal olfaction proceeds apace, robot noses of various kinds are teaching scientists a lot about the billions of smells swirling through the world.

Meet the gas chromatograph–mass spectrometer (GC/MS), a longtime staple of smell technology. Researchers can siphon a sample of air into a GC/MS and reverse-engineer exactly the scent molecules present in it. They can even determine which molecules are most responsible for the scent's dominant notes. Nearly all smells are comprised of many molecules, and sometimes a molecule present in only scanty concentrations turns out to be crucial to the smell we perceive. Fragrance chemists regularly use a GC/MS to parse the "secret" molecules in their competitors' perfumes—this is both legal and fairly accurate. This technology is also used widely in chemical engineering, law enforcement, and environmental cleanup.[30]

Electronic noses are already hard at work detecting telltale smells in various situations. One way to produce an electronic nose is to clone the genes for the four hundred olfactory receptor types humans typically have, and then measure individual responses to each one. These cloned genes can be grafted into living mammalian or yeast cells to create a cell-based e-nose, or what's known as a cell-free sensor. You can think of this as a true robotic nose that doesn't need special conditions to stay alive.[31]

Electronic noses are usually programmed to recognize a few smell molecules specific to their purpose. For instance, a beer brewer could use an e-nose as an early warning sign if beer is oxidizing or developing other unpleasant stinks. (For more on beer's smells, see chapter 7.) Either right now or very soon, e-noses will be able to smell when to pick a rose to optimize its freshness,[32] when to toss spoiled food,[33] and when to diagnose early onset of many diseases, from cancers to COVID-19.[34] More recent advances in e-nose technology are developing sensors capable of detecting many different smells accurately. Think of these as robotic noses with the flexibility and wide-range capabilities of mammalian noses.[35]

Other e-noses are focused less on identifying exact smells and more on sorting smells into groups based on perception, answering questions like: how likely is this smell to be considered unpleasant or intense? Using human subjects and AI algorithms, researchers are also making headway in connecting a smell molecule's shape with what it smells like subjectively. Reverse-engineering a smell based on its molecular structure sounds exciting, but it's in its early days yet. A pioneering 2015 study made headlines with some very broad-sounding takeaways. Researchers learned, for instance, that sulfur atoms make molecules more likely to smell burnt or garlicky, and that bigger molecules are more likely to smell pleasant than smaller ones. The fact that this stuff is headline news only confirms how little we actually know.[36]

The holy grail will be to discover the common denominator among smells, the olfactory equivalent to light wavelengths to measure color or audio frequencies to measure sound. How can you quantify smells so that wildly divergent odors can be directly compared? Many smell classification systems have attempted to boil this particular ocean, sorting billions of smells into objective categories. Perfumers have made some headway for the palette of smells they deal with; scientists are still largely at square one. But history is littered with such attempts.

Naturalist Carolus Linnaeus proposed a scientific taxonomy of smells in 1752 that only addressed floral scents and was widely panned for scientific use. About a century later, the British chemist George William Septimus Piesse compared smells to music with his "odophone" concept. His classification system organized smells in perfume atomizers along a piano keyboard. Piesse paired heavy odors with low musical notes and lighter or sharper scents with higher musical tones. His ideas survive in the language of perfumes still, with perfumers referring to specific smells as "notes" and blending them into "accords." Like music, perfumes take time to reveal themselves fully. First you notice the "top notes" that evaporate quickly, then the "middle" or "heart notes" as the scent develops on your skin, and finally the fading "base note" which lasts the longest.

Later scientists took their own cracks at smell classification, each system more successful and scientifically rigorous than its predecessors. In 2018 a group of scientists decided to organize odors based on how frequently they occur together in nature. They mapped smells onto a curved 3D shape mathematicians call a hyperboloid (picture a Pringles potato chip). Organizing smells in this way revealed interesting affinities: some directions on the map were distinctly pleasant, others unpleasant; some directions correlated with acidity or how easily smells evaporate from a surface. It's impressive stuff, but only because it underscores how wide-open the field of smell actually still is.[37]

Let's wander into that field and see what else we can discover, shall we?

Part Two

THE SMELLS

1 FLOWERY  HERBAL

Exploring smells of the plant world, delving particularly into the history of perfume-making as well as smell's invisible connection to air. You'll start to see how smell represents a weird aeration of everything.

PETRICHOR

ROSE

JASMINE

FRESH-CUT GRASS

LINE-DRIED LAUNDRY

PETRICHOR

Petrichor, the smell of parched earth after rain, is immersive, roomy enough to move around in. Its bright mineral tang is edged with vegetal green. There's a hint of sourness, haloed by fresh water droplets.

Petrichor lifts the ground, with all its smells, closer to the nose. It's as if the earth has exhaled. Because this smell emanates from millions of pinpoints at once, petrichor has a stereoscopic quality. Inside the smell, each moment seems to dilate and slow. It fills the air with relief.

Petrichor is seasonal. You can't smell petrichor whenever you want to; the atmospheric conditions have to be just right. It's not so much a scent as a happening.

Most of us love the smell of dry earth after rain, but don't realize it has a name. (Actually, it has not one name but several interrelated ones.) We've been smelling this scent all our lives, but rarely talking about it. It's everybody's delicious little secret.

In the Indian city of Kannauj, they make a perfume called *miiti attar*. From April through May, perfumers pry blocks of parched clay from the ground. They bake them into discs, then warm the discs over water distillers. When the clay reaches the right temperature, steam slowly releases the earth's smell. The scented steam travels as water vapor through a tube, dropping into a vat of sandalwood oil, the perfume's base.

One message petrichor conveys: you can't smell anything unless that thing is changing in some way. No change, no smell. In Kannauj they're bottling petrichor as the smell of abatement, the scent of the day when months of dry heat give way to monsoon season.[1]

The word *petrichor* means "blood of stones" in Greek. It was coined in 1964 by two Australian scientists, Isabel "Joy" Bear and Richard Thomas, who extracted a golden oil from a variety of soil types: sand, dirt, clay. They learned that plants secrete fatty acids, mostly palmitic and stearic acids, into soil—and that those secretions get concentrated between rainfalls.

Noticing that after a drought ends, plants often have a surge of growth, Bear and Thomas wondered if petrichor was a fertilizer. It wasn't. It turns out that the plants are playing defense. They're secreting the fatty acids to slow the growth of nearby plants, reducing competition when water is scarce. It's this buildup of plant-released chemicals that we smell as petrichor.[2]

A few years ago, a team of MIT scientists figured out how petrichor reaches our noses. Using high-speed cameras, they observed that when a raindrop hits a porous surface, it traps tiny air bubbles at the point of contact. MIT News described the process: "As in a glass of champagne, the bubbles then shoot upward, ultimately bursting from the drop in a fizz of aerosols." Sudden, light rain on sandy soil operates the same way, yielding the best possible aeration to release the petrichor smell.

Online I watch the MIT video in slo-mo as a raindrop hits a surface. The drop briefly assumes a doughnut shape, then flattens into a disk. Infinitesimal droplets rise from the disk, like fireflies buzzing over a lake. Those droplets lift petrichor from the soil, infusing the air.[3]

Rain produces its smells in stages. The first smell is ozone, clean-scented and metallic. Before a thunderstorm hits, air pressure and temperatures fluctuate. The air's turbulence breaks up atmospheric molecules, turning oxygen into ozone. Downdrafts then push the ozone out ahead of the storm.[4]

Petrichor marks rain's actual onset. As the rain keeps falling, that smell eventually deepens into a related but different smell, geosmin. This is the musty, wet scent that lingers after rain ends. Related to the smells of beets and of freshly plowed earth, geosmin arises from busy soil bacteria called actinomycetes. Geosmin might exist to tell freshwater fish where to spawn, or to guide camels to replenished desert oases.[5]

Petrichor emphasizes another idea: smell is what happens when a substance transmogrifies into air. In the case of petrichor, that transmogrified substance is a golden oil, secreted into soil by rivalrous plants. Smells ride on air, impart a kind of personality to it.

Petrichor also reminds you that air is three-dimensional; it takes up space. The air's stillness after a rain shower is monumental, pungent, and temporary. I picture petrichor like a golden cube of air above the ground, trembling like amber-scented Jell-O. As you smell anything, but especially when catching a whiff of petrichor, you're observing a change happening *here* and *now*. Each smell pins a moment in space and time—with you, the smeller, as its witness.

On a frigid Chicago day, I dab *miiti attar* on my wrists, step outside, and sniff. The petrichor smells charged and potent. Here, there are still weeks of hard winter to go, but petrichor's headlong smell brassily pushes that calendar forward. It smells of incautious hope, of barreling into spring.

ROSE

It's so iconic, this smell, so recognizably itself that it takes time to actually observe its details. At first the smell of roses fills the nose in a giddy rush, sweet and headlong. The burst of luxury feels total. But that rush rapidly settles down and shades into a regal stillness. You become aware of joining a vast throng of admirers in a shared contemplation.

The scent's wildness doesn't stay wild for long in the nose; inevitably, one recalls one's grandmother in her church best. Yet behind its rounded, classical quality, the scent still emanates some heat, prickly and pollen-like. It evokes late summer, lazily crawling insects, the finery of silken petals rotting under a bush, the sprawl of untidy nature. Inside the smell of roses one senses the momentary quality of life, how rapidly loveliness blossoms and fades, the nearness of beauty to rot.

Researching the history and symbolism of roses is tricky. It seems these flowers intoxicate us more than anything with an excitable vagueness. Roses symbolize so many positive things—love, sex, spirit, magic—one wonders how far the symbol can stretch. What *can't* roses mean? "Prosaic" perhaps, although their ubiquity in art, architecture, textiles, and fashion undercuts this point. Roses aren't usually praised for their discipline or rationality. Nearly all cultures and historical contexts have found expression for their most exalted concepts in roses. It's flower royalty.

Dabbing oneself with the scent of roses, as modern Westerners do, is probably the most tepid way of communing with this smell. Diehard fans also eat roses, drink roses, consume roses as medicine. They perfume their own bodies, whole crowds of friends, shared spaces, the entire great outdoors if they could. Inhaling the smell of roses provokes for some people a delirious and wildly expensive thought: why can't *all* air smell this way?

IT'S HARD TO SUMMARIZE EVERYTHING ROSES— especially their scent—seemed to have symbolized to the ancient Romans. "Life force" is probably the best approximation. Wealthy and powerful Romans showered roses on bridal parties and winning athletes alike. They'd adorn guests with rose wreaths, sprinkle rose water over them, strew rose petals all over dining tables, over the food, set rose petals swimming in wine. One famous dinner featured rose-water-drenched doves who flew around the room dousing the guests with fragrance. The Emperor Nero outfitted his dining room ceiling with a contraption that enabled him to shower rose petals and fragrance down on guests between courses. Another Roman emperor, Heliogabulus, accidentally suffocated his dinner guests in an avalanche of rose petals.[1] At outdoor sporting events, Romans anointed the awnings shading spectators, so they could watch and recline surrounded by rose-redolent air. The springtime celebration of Rosalia honored the dead with out-pourings of roses and violets. To the Romans, a cut rose resembled a beautiful youth, struck down suddenly in his or her prime by death.[2]

In the Muslim world, particularly in Persia, the scent of roses unites spiritual and earthly love. It encompasses sex and romance but also transcends them. In the *Hadith*, the collected sayings of the Prophet Mohammed, when the prophet ascended into heaven to receive his divine revelation, sweat dropped from his brow and penetrated the earth below, sprouting forth as fragrant roses.[3] A common motif in Persian art pairs the nightingale singing his song to a rose, the two separated by a high wall. The theme suggests to me several overlapping interpretations: a lover separated from his beloved, or a believer yearning for God. Their bond feels even more evocative because it operates blindly, beyond the visible: only the audible song and the wafting scent connect them to each other.[4] This invisible bond is not dissimilar to the actual, scent-based bond be-tween roses and the insects that pollinate them. These insects will tunnel toward a rose's scent even with extraordinary obstacles in their way: powering through the blasts of a wind tunnel, sensing flowers that are trapped hermetically under glass.[5]

The enthusiasm for roses extended to the Mughal Dynasty, founded in 1526 by Zahir ud-Din Muhammad who's better known as Babur. Following Babur's lead, formal gardens stuffed with roses and narcissus flourished all across Persia and India. Babur named his three daughters for roses: Gulrang (Rose-Colored), Gulchihra (Rose-Faced), and Gulbadan (Rose-Bodied). The Persian word for rose, *gul*, is also that language's generic word for flower, signaling the rose's pre-eminence in its category.[6]

Across the Muslim world, roses unfurled everywhere, embodying a kind of live and fleeting beauty. In Persia and Damascus, people would bury unopened rosebuds underground to store them in their nascent state. For important dinners they'd dig up the rosebuds as decorations. When heated with the food, the rosebuds would flare dra-matically open on the plate.[7] Arabs often mixed rose water into the mortar used to build mosques, so the walls would exude fragrance on hot days.[8] Many Middle Eastern sweets are still flavored with rose water: barfi, baklava, halva, gulab jamun, kanafeh, nougat,

Turkish delight. In the Indian subcontinent a soda popular at Ramadan, Rooh Afza, is flavored with rose water. Roses are ubiquitous but also somehow special. Their presence exalts ordinary life, if only for a moment.

Back in Europe, classical Romans' rose-delirium spread across the continent even after their empire fell. The Christian Church attempted to suppress roses as pagan and sensual but ultimately failed. Instead they associated the Virgin Mary with roses, channeling a secular enthusiasm into worship. Christian rosaries were originally made from actual rosebuds. Fingering each bud released a puff of scent, helping the believer enter into the reverie of prayerful meditation.

Through the centuries, roses continued their association with momentary exaltation, with sensuality and liveness that cannot last forever. Wealthy Elizabethans hurled eggshells full of rose water at one another as a scented amusement at dinner parties. Roses scented lots of luxury objects in Baroque seventeenth-century Europe: notepaper, linen, bathwater, lamps, shoes, and jewelry—bracelets might include aromatic beads one could finger to release scent, and rings were designed to squirt a jet of perfume. During this period, smell was regarded as evidence of an object's true essence, akin to its soul. The look of things often contrasted with their insides; vision could be duplicitous.[9] Smell was considered the realest sense, if only briefly, and roses arguably the most beautiful expression of that sense.

IN THE PAST AS WELL AS THE PRESENT, perfumers love working with roses for many reasons. Roses come in many fragrant varieties, especially older varieties like Alba, Gallica, Centifolia, and Damask types. As a perfume ingredient, rose always plays well with other scents. It's also forgiving: adding a little rose essence can fix a wayward perfume mix and soften mistakes.[10]

Breeding roses is such an extensive and fabled activity, the scientists and gardeners who specialize in it are known as "rosarians." You can easily trace the history of East meeting West in the increasing—and sometimes bewildering—entanglement of rose strains. Many resulting rose varietals are themselves personalities, named for famous people: Dolly Parton, Princess Diana. Quite a few rose names celebrate people who might otherwise be lost to history: for instance, Madame Testout, a nineteenth-century dress designer in Paris. Roses cannot escape celebrity, it seems.

What has escaped many roses, of late, is their scent. Many of the most popular varieties nowadays are bred for visual showiness, color, and shape, not for fragrance. But scientists have discovered the enzyme in roses that imparts its sweet fragrance. The RhNUDX1 enzyme, active in the flower's petals, produces monoterpene geraniol, the primary chemical in rose oil. They hope to breed this enzyme's genes back into modern, odorless rose varieties to bring back the scent.[11]

There's something sad and elegiac, almost impossible, about re-engineering the fragrant soul back into roses. If roses remind us how temporary beauty is, their scent reminds us this beauty is, like us, also alive. Smelling a rose's scent can pin our own temporal and physical selves in a particular moment. Here we are, sniffing, enjoying, mortal. It's an exalted and quiet experience, abundantly available until suddenly it's not.

JASMINE

My jasmine perfume rubs waxy and butter-yellow onto the skin. A moment of melt, and it seems to disappear. Then comes an extraordinary sense of liftoff. The scent rolls over you like a beautiful tidal wave, cresting as it first breaks. So! much! flower! One buzzes around and around inside the smell, marveling at its building heat, its texture both powdery and silken-fine, its sense of a deepening golden bower.

The top notes fade rapidly and reveal a warm, resonant middle note. Now the beauty is both ample and gracious, unhurried. It captures all the loveliest qualities of home: a lovely place that's also eminently comfortable. One slips into jasmine like a second skin.

You can drink jasmine tea or eat jasmine rice and thus indirectly enjoy its scent. But mostly jasmine is synonymous with perfumes—"no perfume without jasmine," as the perfumers' saying goes. Its scent contains an entire world of notes, uniting the sublime with the everyday.

Perfumers like jasmine for many reasons. It's a miniature perfume on its own, complex and balanced, and elevates almost any other scent it's combined with. In perfume accords or mixtures, jasmine is a middle or "heart" note, lasting a relatively long time. Its scent also combines both fragrant and foul notes in an intriguing way. Its honeyed, warm sweetness is edged with a dash of shit and decay, thanks to its high concentration of the molecules indole (also in feces) and cresol (that sweetly nasty tinge of coal tar).[1]

The jasmine plant comes in two main varieties: *Jasminum grandflorium*, or "royal jasmine," and *Jasminum sandac*. (Another variety, *Jasminum vulgaris*, smells less sweet but is useful as a root stock for grafting royal jasmine. And there are dozens of other flowering vines colloquially called jasmine but unrelated to these species.) Jasmine is most closely associated with the town of Grasse in southern France, the historical center of French perfumery since the sixteenth century.[2] But the flower is native to India and grows lushly in many widely dispersed regions—Iran, Turkey, China, Morocco, Egypt, Bulgaria.

Jasmine is a climbing vine with small white blossoms, unassuming to the eye. Indians refer to jasmine as "moonlight of the grove" because it blooms only at night, releasing its fragrance under cover of darkness and leading assertively with its scent.

Jasmine's scent is difficult to imitate perfectly with synthetics, so natural jasmine is preferred for perfume. It must be hand-harvested in the early morning hours, while the blooms are still open and scented oils reach their peak levels.

For most of history, steam distillation has been a workhorse process to extract floral scents for perfume. Dating back to 3500 BCE,[3] steam distillation traditionally happens in a pot-bellied vessel with a thin, coiled tube snaking from its top down to another container. It begins with plunging flowers into boiling water inside the closed vessel. This releases scented oils into the gathering steam. The oil-laden steam rises upward and enters the coiled loop, which cools the steam down as it travels down the tube. The steam turns back into liquid water, now with a layer of scented oil floating on top.[4] One can repeat the distillation process to refine perfume oils further.

But for especially delicate flowers like jasmine, steam distillation destroys the very scent molecules it seeks to capture. In premodern times, perfumers captured jasmine oils with a process called *enfleurage* instead. The perfumers would choose a non-evaporating fat—like oil of sesame, olive, almond, or mastic—or else refine an animal fat like lard or tallow to make it odorless. They'd smear the fat onto glass plates set into wooden frames, then press the blossoms into the fat. The flowers' scented oils would slowly release into the fat. The perfumers would then keep adding fresh flowers until the fat was fully saturated with scent. Enfleurage suits flowers like jasmine because they release different concentrations of oils in the hours after they've picked. Enfleurage works like a time-lapsed photograph, capturing a long, slow picture of the flower's nuanced scent.[5]

Distillation is beautiful. It is a slow, philosophic and silent occupation . . . [Via] metamorphosis from liquid to vapor (invisible) and from this once again to liquid . . . purity is attained, an ambiguous and fascinating condition which starts with chemistry but goes very far.

—Primo Levi, *The Periodic Table*

In the modern era, extraction is the preferred method for capturing jasmine's scent. Blossoms are strewn on perforated sheets of metal, then lowered into a sealed vat where a solvent (usually hexane) circulates around them and heats the flowers up. The solvent dissolves the scented oils, waxes, and pigments inside the plants. Evaporating the solvent creates what's known as a concrete, a waxy solid perfume. Concretes can be worn as a perfume themselves, with a sparing touch. The jasmine I'm wearing now arrived as a tiny, dense smear of concrete perfume in a round container like a locket. Or you can dissolve some concrete perfume into alcohol and filter it to produce what perfumers call an absolute.[6] Even using the most efficient extraction method, it still takes three thousand pounds of jasmine flowers to produce two pounds of jasmine absolute.[7]

Stoppered in a crystal bottle, jasmine perfume is clearly an ultra-luxury. But it's also an everyman perfume in the wild. Jasmine is a humble little wildflower that grows abundantly in many places; in those places its smell carries with the wind, a simple and ubiquitous pleasure.

In Beirut and many Mediterranean towns, people regularly plant jasmine over doorways to welcome visitors with a whiff of fragrant, fresh air. Mothers leave plates of jasmine flowers by their sleeping children to sweeten their dreams. People make jasmine-bud necklaces to sell; these are especially popular among taxi drivers. The buds bloom fragrantly into full flowers around the wearer's neck, then drop off as the scent fades. In many places, especially India and Lebanon, brides wear a crown of jasmine to symbolize domestic happiness. In Lebanon, when you wish someone good morning, you say *sabah il-khahr,* to which they might reply the same or intensify the wish: *sabah il-nour*, "morning of light." An extra-warm greeting to loved ones invokes the flowering jasmine: *sabah il-fil wal yasmin*, "I wish you a fragrant morning full of goodness, like the jasmine."[8] At home around the globe, jasmine abandons haute luxury and instead becomes friendly and hospitable.

I usually don't wear any perfume, so wearing jasmine repeatedly was an unfamiliar habit. But jasmine got me thinking and learning about perfume and wore down my reluctance to engage with it. I was intrigued by some basic but surprising facts. First, perfumes smell wildly different on different people. You cannot judge a perfume by sniffing the bottle, or a *touche* (those paper strips popular at perfume counters), or sniffing someone else's wrist. Perfume is alchemical with *your* skin. It registers subtly differently across ethnic groups.[9] It even registers differently across zones of the body:

perfume daubed on your dominant wrist will smell more potently than perfume on your non-dominant wrist. Perfuming one's temples or hair really boosts the scent thanks to your scalp's oil glands. (This also explains why wealthy ancient Romans anointed different pulse-points of their body with different scents.) Perfume mixes with, but also amplifies, your thrumming and individual body.

Second, perfume changes over time and is intentionally designed around this fact. The first moment of contact with the skin can feel showy and impersonal, revealing all the perfume's angles at once. But then the scents meld into the skin and shape-shift. Perfumers often describe a perfume's structure using musical metaphors: The top notes are most volatile and dissipate first. Then the middle or heart note surges forth. When that departs the base note lingers, like a ghostly coda of the entire composition. Music is aleatory and time-based, too.

Third, perfume comes alive with movement. I loved learning the French word *sillage*, which describes both a perfume's staying power and the physical trail of scent it leaves behind a wearer. You cannot smell your own sillage, of course, but wearing perfume does fill the day with unexpected jolts of loveliness. I'd brush a fly from my face or hurriedly push up my sunglasses, and the movement would strobe jasmine's scent from my wrist back into my nose. It was startling in the best way.

Wearing jasmine daily got me thinking differently about perfume as a category. Yes, it can signal wealth and prestige in occasionally crass ways. Yes, historically it's anti-feminist. Yes, it's irritatingly French for a non-Francophile like me. And yet, thanks to jasmine's friendly and unassuming proximity, I can also now appreciate perfume as a private pleasure. It can make your personal style multisensory. It makes you feel protected, warmed, ready to enter the public sphere. If clothing is armor, so, too, is perfume. And when you dismantle your armor and fully reveal yourself, the perfume you wear still clings to your naked body.

Jasmine is probably not "my" perfume, but it signaled a new possibility: finding my perfume could be like finding my best friend. Somewhere out there is a scent that will click resoundingly with me. Now I relish the search.

FRESH-CUT GRASS

It smells juicy, fresh, and green, especially concentrated near to the ground. The scent makes you feel lifted up like a wilting plant drinking in water. Sniffing fresh-cut grass feels intimate. With every inhale, your lungs are filling with the exhalations of plants.

I could smell it keenly while I mowed. As the blade whirred and the shorn grass arced upward in crazy tufts, I fell into the scent's slipstream. It mixes pleasantly with other scents: my own sweat mixed with sunscreen, puffs of summer air drifting by, the heavy undertow of garden tomato vines, the humid moss beds and ferns in the yard's deeper shade. Yet it's shudderingly brief, this smell. As soon as I stopped mowing, the scent of cut grass tailed off rapidly. Cutting the grass creates a private and temporary summer bower, a bubble that moves with you only as long as you work. When you're done, the mood snaps and the smell vanishes.

I gathered some clippings into a plastic bag, which immediately fogged up with moisture. Handling fistfuls of grass is really satisfying, tactilely speaking. You're tempted to paw and paw at it, turning it over in your fingers like a tiny haystack. It's abundant, pliable, even pillowy. But smelling the grass itself was not at all like smelling the mowing as it happened. The grass itself smelled more earthbound, loamy, faintly sweaty. It's the straightforward odor of plants, no longer the living out-breath of botany.

It's no accident that fresh-cut grass smells like draining vitality. With this odor, the blades of grass are signaling madly to each other that danger is approaching.

Like most plants, blades of grass communicate with each other via smell. It's well known that flowering plants attract pollinators with their smells, and fruiting plants use scent to lure animals into dispersing their seeds. What's less well known is that plants talk to each other, a lot. Unlike animals, plants are rooted to a single spot—they can't move to evade their predators. What they can do is prepare other evasive maneuvers and warn each other via smells.

When under attack, plants can release volatile compounds to warn other plants of

EXERCISE

Start a smells journal.

Improving your sense of smell begins with activating your powers of observation.

Think fast: so far today, have you noticed *any* smells? The answer might well be no, and that's fine. Try to find one smell today and stop to notice it. It's fine to seek out the most obvious smell sources you can think of. Remember the last time you strolled past the mall's Cinnabon or wandered by the perfume counter at a department store? Simply walking through your backyard on a warm day after rain should give you good scents to observe. Jot down your smell in a notebook or a text doc on your phone, along with today's date.

As you repeat this habit, two things will start to happen. First, you'll notice smells more readily and jot them down with greater frequency. Second, over time this habit will build a surprisingly evocative diary, your life captured in moments and smells.

the impending threat. The warned plants can get busy, rapidly making parts of themselves less nutritious or tasty to predators or getting ready to heal potential damage. One part of a plant might even signal to other parts of itself that attackers are near. Plants can even send olfactory signals to predators of whatever insect is attacking them, on the logic of "the enemy of my enemy is my friend."

When grass is damaged, it activates an enzyme called lipoxygenase and unleashes two acids. These acids, now exposed to the air's oxygen by the damage done, are transformed by a different enzyme into (z)-3-hexanal. This is the keen olfactory edge of the grass' suffering, the fresh-cut grass smell that humans so thoroughly and heartlessly enjoy. But (z)-3-hexanal doesn't stick around long in its complete form. Like an urgent telegram received and then crumpled into the trash, (z)-3-hexanal breaks into smaller molecules and then dissipates rapidly from there.[1]

The way plants talk through smells isn't well understood, but apparently it goes deep. Belowground, tree roots are interconnected by vast and dense fungi networks nicknamed the "Wood Wide Web." This tangled network enables trees to share nutrients with offspring or neighbors, to warn each other of impending attacks, or even to sabotage their competition. They do all this via chemical signals, to which only trees and plants are attuned.[2]

Plants are always semaphoring invisibly to their fellows. Catching their smells is a way of fluttering the curtain on a kind of hidden communication. We may find the smell of fresh-cut grass lovely, but to the grass itself this odor is an urgent communiqué, an SOS. What we humans think of the smell is unimportant; it's not actually meant for us.

LINE-DRIED LAUNDRY

It smells like the entirety of outdoors, shrunken down in scale. The smell brings this sense of largeness with it, imbuing everyday cloth with whiffs suggesting enormity. Its scent hides in the crenellated folds of your underwear. It tucks into your bed along with the fresh bedsheets. Even though your feet are too far away for you to smell, the scent of line-dried laundry wreathes your clean socks, too.

It's sun-washed and warm, bringing the afterglow of a late summer's day indoors. The actual scent feels vaulted with many ethereal top notes, real but ghostly. Its base note is clean fabric, unadorned and humble. Breathing this scent in, one pictures the wild aerial twistings of cloth amid gusts and lulls, a freeform semaphoring of a domestic flag made possible only by a well-tethered clothesline. You want to bury yourself in this smell nose-first.

How could you study such a smell scientifically? An atmospheric chemist who grew up in laundry-littered Italy ran a simple experiment to find out. Along with two fellow grad students and her adviser, she washed three cotton IKEA towels and hung them to dry in three configurations: one indoors, one outdoors but shaded by a tarp, and the third outdoors in full sun. Once fully dry the students measured the smell molecules released by each towel and compared them to one another, as well as samplings from an empty bag, an unwashed towel, and the air circling around the drying sites.

By comparing and contrasting all these air samples, the researchers isolated the smell molecules coaxed out of laundry specifically by sunshine. These included many aldehydes and ketones including pentanal (found in cilantro and many alcoholic beverages), octanal (which gives off a clean, citrusy odor), and nonanal (a scent balanced between roses and melons or cucumbers). The students formed theories: it might be atmospheric ozone that converts common chemicals in the laundry detergent and cloth into aldehydes and ketones.

Or it might be the sunshine itself performing the alchemy. Exposure to ultraviolet light excites certain molecules into becoming highly reactive radicals. In turn, these radicals react with other molecules to produce an excess of aromatic ketones, aldehydes, and esters. These three molecular families are associated with good natural smells in plants, foods, and perfumes. The researchers wondered if water in the wet towels might further concentrate the sunlight's rays like a magnifying glass, accelerating the molecular reactions and coaxing forth even more of this fragrance.[1]

Line-drying laundry outdoors is a strangely embattled practice in America particularly—so much so that lawyer Alexander Lee cofounded Project Laundry List dedicated to defending the right to line-dry laundry in public. The nonprofit's argument for line-drying is chiefly environmental but not exclusively. "We are not anti-dryer; we are pro-clothesline," Lee tartly observed in one magazine interview. The real problem, he notes, "is the hundreds of millions of Americans who refuse to get up, go outside for some fresh air and sunshine, talk over the fence with their neighbors, and mindfully take time to do an essential human task. By my estimate five billion plus people in the world manage fine without a dryer. It may not be 'easy living,' but it beats having the ocean lapping at your door."[2]

A lot of people think clotheslines look low-class and complain about them on NIMBY grounds. But line-drying your laundry is beautiful and slow. It stretches time, breaks the summer afternoons into manageable chunks of time. Time to switch out the laundry. Time to fold it and then replace it in dark drawers. And line-drying suffuses a kind of everyday, aromatic magic into your clothes.

People who think line-drying clothes in inner cities would make your garments stink are dead wrong. In Berlin where we spend most summers, dryers are rare and everyone line-dries their clothes all year long, indoors or outdoors. Our clothes hang over an exhaust-filled street with pedestrians below smoking cigarettes and eating currywurst and spilling crap beer on the sidewalk and pissing semi-discreetly in street corners. Berlin is pretty filthy at ground level. But just two stories up, the air is entirely different.

My husband, Seth, ribs me about my zeal for doing laundry, particularly in Berlin. My protests seem levelheaded enough to me: the washing machines there are tiny. The supply of clothing we bring on these trips is finite, and usually half of it is totally wrong for Berlin's unpredictable weather. We spend lots of time outdoors there: biking everywhere, sweating through A/C-free heat waves, romping in sand-filled playgrounds. It's a more analogue way of living that soils your clothes much faster.

Laundry in Berlin is necessarily a daily ritual. Line-drying our clothes isn't a romantic or idealistic choice; there it's pragmatic. There's no other way. But the practice does mark a gulf between ordinary life in Chicago versus ordinary life in Berlin. It's a domestic rondelay I enjoy thoroughly, and all these yearnings and ideas seem to permeate our clothes as a scent.

In Berlin my laundry detergent of choice is Persil, a brand loyalty every bit as unshakeable as my loyalty to Tide in the States. I'm not alone among consumers who feel strongly about their preferred laundry detergent. It sets your baseline expectations of what "clean" smells like, made more potent by the fact that it may've been chosen by our mothers, or even grandmothers. My mother—both scrupulously clean and a fantastically thorough caregiver—washed our clothes with Tide throughout my childhood, and I have zero interest in altering that pattern.

My friend Andaiye told me two stories about laundry smells that have stuck with me. In the first story, she remembers walking down a narrow cobblestone street in Toledo, Spain, which was festooned with billowing laundry lines. This was her first trip to Europe, and she took in big lungfuls of the great-smelling air as she walked under the crisscrossing lines overhead. Later on, thousands of miles distant, she smelled that exact scent in Newark and felt instantly transported to the Spanish alley.

The second story was more quotidian: less positive anecdote and more simple reminiscing. Both of us grew up with line-dried laundry, so we spent a minute recalling how laundry lines create their own outdoor stage setting, like a movable fort. We recalled dashing between flapping wet bed sheets, how the sunlight dappled through rising and falling cloth and changed with the passing hours, the basket of clothespins with their particular quiddity. Both stories hung on laundry smells but also on ghosts and time travel, both lingering invisibly in the air and inside folds of clothing.

I'm always a sucker for stories of how smells foreground air and make it briefly visible, teeming with unseen activity. Laundry hung out on a line to dry does that. It swells with the outdoor air and fills with its surroundings' smells, a slice of time, space, and weather.

2 SWEET

An armchair tour of beloved foodstuffs, which both deepens and complicates what these familiar sweet scents connote.

VANILLA

SWEET WOODRUFF

BITTER ALMONDS

CINNAMON

HOT CHOCOLATE

VANILLA

The scent is languorous, deep, and sweet. Smelling it makes you feel like you're resting in a still, heat-filled shadow that pools and pools around you. It smells replete.

Vanilla's scent also, incongruously, makes you feel talky. And no wonder, because this smell appears in almost every tasty treat. Sniffing it repeatedly starts an ever-expanding game of free association. You can sniff for various dried fruits and find them each in turn: raisins, dates, prunes, currants, apricots. Sniff again, and the scent shape-shifts between granulated sugar, aged wood, hothouse flowers, cured tobacco. You can find the deep, syrupy liquid of brown liquors. All the while the mind clicks through a pleasant visual slideshow, cycling through every possible shade of amber.

But vanilla's tantalizing smell hides an important fact: it's knowable only indirectly. Vanilla reminds you effortlessly of its many companion smells and flavors, yet smelling "just vanilla" is almost impossible. Vanilla is now synonymous with blandness but only because of its universal popularity and its ever-supporting role for bolder flavors. Because vanilla now flavors almost everything, it fades into the background but also meshes intriguingly with it.

Its whiteness seems to hold a clue. But natural vanilla isn't white at all; it's black. Specifically, the aroma comes from black "caviar", seeds scraped carefully from the insides of equally black vanilla beans. These beans come from one of three species of edible orchids: *Vanilla planifolia* and *Vanilla pompona Schiede* are native to southeastern Mexico and Central America and later were transported to Madagascar, Tahiti, and Indonesia.[1] A third edible species, *Vanilla tahitensis*, is a natural hybrid of the other two.[2]

Vanilla's history has always been closely intertwined with chocolate; flavors we now see as white and brown, light and dark, are long-standing companions. The Totonac people in pre-Columbian Mesoamerica are credited with first domesticating the vanilla plant, paying vanilla beans in tribute to the ruling Aztecs. They called vanilla *caxixanath* (catch-ee-CHA-nat), or "hidden flower," because it wasn't particularly showy and its blooms wilted within hours.

The Aztecs added vanilla beans to soften the bite of *chocolatl*, a bitter hot cocoa drink well known as an aphrodisiac and often served on formal occasions.[3] (For more on chocolate's history, see the hot chocolate section in chapter 2.) Moctezuma served it to Hernando Cortés in 1520, ushering vanilla into the West's culinary ambit. The Spanish renamed caxixanath *vainilla*, a diminutive for *vaina*, Spanish for "seed pods."[4]

Until the mid-nineteenth century, vanilla beans were the product of luck followed by intensive labor. Like all beans, vanilla grows on vines which must grow around a "tutor" tree for support. It can take up to three years for the first blooms to appear. When they finally do, the flower survives only a few hours. Odorless to humans but delectably fragrant to certain bees and insects, the fragile white blooms must be pollinated during this window before they shrivel up.

If luck holds and pollination happens, it takes up to nine months for a seed pod to mature. Once it's harvested, the seed pod must be cured in a four-stage process that lasts anywhere from three to nine additional months. First the plucked beans are "killed": the beans are wrapped protectively in hessian cloth, and farmers then dip the beans briefly in near-boiling water. This stops photosynthesis—"killing" the bean—and jump-starts an enzyme that produces vanillin, the prized aroma molecule in vanilla. Immediately after killing, the wet beans are wrapped in sacking to stay warm and sweat for twenty-four hours. They cannot stay wet, however, or they'll rot. The next phase, drying, involves exposing the vanilla beans to direct sunlight during the morning hours over several weeks, weather permitting. In the final phase, conditioning or curing, the beans are sorted by size and stored in a dark, dry place. Slowly the vanilla beans darken, developing a pliable, leathery texture.[5] Amazingly, freshly plucked vanilla beans smell of nothing. It's only these months of coddling that finally coaxes the aroma out.[6]

Wealthy Europeans became smitten with the twin flavors of vanilla and chocolate spirited over the Atlantic Ocean. The Spanish effectively cornered these markets for many years, but the French and Portuguese hoped to cultivate vanilla in their own tropical colonies. Those attempts failed, since pollination required insects only found back in the Americas.

It took an observant young Black man to discover how to hand-pollinate vanilla flowers and thus found a global industry. Edmond was a twelve-year-old slave in La Réunion, an island off Madagascar, whose white owner, Ferréol Bellier-Beaumont, occasionally tutored him in botany. Bellier-Beaumont had been growing vanilla plants on La Réunion for twenty years, without yielding a single seed pod. One day in 1841 he noticed not one but *two* pods had sprouted—and not, as it turns out, by accident. Edmond had hand-pollinated the flower and demonstrated how he did so. Using a thin stick (or a thumbnail), he lifted the rostellum—the flap separating the male anther from the female stigma—and smeared the pollen from the anther directly onto the stigma.[7] Delighted, Bellier-Beaumont invited neighboring plantation owners to learn *la geste d'Edmond* and pollinate their own vanilla flowers.[8]

This history is, like both chocolate and vanilla in their natural state, more bitter than sweet. Anticipating emancipation in 1848, Bellier-Beaumont freed Edmond early and gave him the curious last name of Albius, meaning "white." With his former owner's support, Albius fended off the false claims of another white French scientist, who said *he'd* taught Edmond the hand-pollination method. Although it must've been sweet to earn historical credit, Edmond Albius's story ends bitterly: neither he nor his descendants benefited financially from his discovery, and Edmond died destitute.[9]

Back in the New World, it's arguable that Thomas Jefferson single-handedly turned vanilla white for generations to come. A Francophile, gourmand, and slave owner himself, President Jefferson returned from France in 1789 with a personal chef well-versed in the latest haute-cuisine fad, frozen desserts. In 1791, amid the French Revolution, Jefferson managed to secure fifty vanilla beans from France. (Note the irony of seeking vanilla beans from a war-torn republic across the ocean, when Jefferson could probably have sourced them from his own continent, one of the places where vanilla originated.)

Jefferson's vanilla ice cream—the recipe for which is stored with his papers at the Library of Congress—was an egg-yellowed white, speckled with black vanilla caviar.[10] Compared with other ice creams of the day—stuffed with nuts, fruit, or even (oddly) brown bread—Jefferson's ice cream was minimalist, even shockingly plain. It began vanilla's fate as basic: as Amanda Fortini writes in *Slate*, vanilla is "merely the starting point for flavor, not flavor itself."[11] It's notable that adding vanilla to chilled ice cream deadens the scent, rendering this aromatic, sensual bean from the tropics odorless.

A perfect storm of historical factors helped vanilla bulldoze the world with its whiteness. Jefferson made vanilla ice cream iconic, at the forefront of a frozen desserts trend among the colonial American elite. In 1841, Edmond Albius's discovery of hand-pollination helped the French dominate global trade in natural vanilla, which ultimately peaked around 1898. Vanilla plantations pumped out vanilla beans to every well-funded kitchen on the planet, not to mention hundreds of hot-chocolate and coffee houses serving vanilla-infused baked goods across Europe. In 1875 vanillin became one of the first flavor molecules synthesized in a lab, widening access to this desirable flavor even more.

The early twentieth century saw vanilla ice cream consumption skyrocket with the rise of soda-fountain parlors. During Prohibition, many bars were converted into ice cream parlors—where vanilla extract, high in alcohol content and thus strictly regulated, might be available for clandestine purchase. This secret tipple makes vanilla once again heavily aromatic, evoking brown liquors forbidden at that time. Soda-fountain parlors persisted into mid-century America as wholesome sites of community gathering but were racially segregated until the Civil Rights era. Soda fountains also provided a marketing platform for launching Coca-Cola, introduced in 1886; Coke's brown sweetness is laced with vanilla.[12]

Meanwhile, generations of housewives of all colors were encouraged to add a dash of fragrant vanilla to just about everything they baked. With the late-century wave of processed and low-sugar foods, manufacturers favored vanilla as a cheap, universally popular flavoring to add nuance to any snack's taste. No wonder vanilla now feels unexciting; maybe we're just vanilla'd out.

And yet, not quite. Despite rampant overuse vanilla still beckons with its quiet charms. It's unusual among smells in that it doesn't activate the trigeminal nerves in the face and nose at all. This increases vanilla's likeability by removing a common objection to smells. Often what we dislike about strong smells is not the smell itself, but the trigeminal response that particular smell triggers.[13]

The scent of vanilla has been found to ease muscle pain, reduce stress headaches,[14] and quell anxiety during MRIs.[15] Aromatherapists claim sniffing vanilla can calm the startle reflexes, ease sleep apnea, and help with male impotence.[16] Its appeal is shy and friendly but very real.

So how did black vanilla become universally known as a white flavor? NPR's *Code Switch* considered this question and offered a complex but trenchant theory. "Whiteness has always, always been defined in proximity to blackness. 'Whiteness is also associated with cleanliness, purity, but also blankness—the lack of color. So I think these ideas are kind of paralleled, the white versus colorful—"colored"—and the chocolate versus plain vanilla,' says Harryette Mullen, a poet and professor who teaches English and African-American studies at the University of California, Los Angeles. 'So it's a way of reversing the kind of implied superiority of whiteness by saying that whiteness is the less interesting color . . . because it's maintained as a norm. And we also have some ideas of how normal is desired but also boring.' "[17] As the default flavor, vanilla's qualities have disappeared like a polar bear cavorting in a blizzard. But like whiteness in other forms, it's time to interrogate that default.

Arraying my three vanilla beans on a counter, I can sniff vanilla from every corner of the globe where it flourishes. The three wrinkled black beans differ a lot in size, and each smells a bit different. The Mexican bean smells most strongly of wood; close your eyes and you picture a cigar-smoking parlor lined with cedar. The bean from Madagascar casts a powdery, hypnotic veil of sweetness. The Tahitian bean is brightly floral; with the lowest concentrations of vanillin molecules, Tahitian vanilla reveals the smell's notes most readily. Vanilla re-emerges, foregrounded and crackling with specificity again.

SWEET WOODRUFF

It combines a few simple notes winningly: fresh-mown hay, a grassy greenness, with hints of almond, vanilla, powdered sugar. It's a bashful smell, not easily coaxed out. For me, this scent is bound up in my chase to locate the smell, a kind of delightful hide-and-seek.

When I started researching an article about sweet woodruff for *The Art of Eating*, I hadn't anticipated how much its taste would depend on its delicate, evanescent smell. That smell evokes for Germans pungent memories of childhood and a rich culinary tradition—memories I lack from my American childhood but found fascinating to backfill in imagination. The article also turned into a thought experiment: how could I possibly recognize a flavor I'd never encountered before? The fake version of this flavor is ubiquitous in Germany. My challenge was to re-create this flavor from the original plant, without knowing what that original should taste like. Rereading the article now, I see the ghostly outline of scent rising, concentrating, in the air and in my mind.

SWEET WOODRUFF, OR *WALDMEISTER* IN GERMAN, translates literally to "master of the woods." The plant has earned equally florid names in English: it's also known as "wild baby's breath" and "sweet-scented straw." The flavor is linked to the warming weather of spring, since waldmeister first appears in the woods in late April and must be harvested before it blooms in June. It's a key ingredient in *Maibowle*, a punch that Germans drink in May and early summer, a heady mix of sweet white wine and *Sekt* (sparkling wine) sweetened with brown sugar, in which fresh sprigs of mint, lemon balm, and waldmeister are steeped. This *Parfumierung*, as the Germans describe it, gives the punch a delicate springlike flavor of fresh-mown hay. It also tinges the drink a pale green color.

I spend a lot of time in Berlin where practically every sweet comes in waldmeister flavor. In spotlessly bright white cases of ice cream around the city, you can always spot the unnatural, livid green of waldmeister. It's drizzled into *Berliner Weisse*, a cloudy wheat beer served in summer, into which violently false green *Waldmeistersirup* is poured. Any green lollipop, gummi bear, or hard candy in Germany isn't flavored lime as Americans expect, but waldmeister. Why?

Turns out waldmeister runs deep for Germans, much deeper than lime flavor runs for most Americans. As a spring ritual, Maibowle dates to 854 CE when a Benedictine monk Wandalbertus of Prünn mixed the first Maibowle from waldmeister, black currants, and creeping ivy.[1] To Germans, waldmeister's scent is redolent of springtime, incipience, greenness. Judging from its popularity in children's treats, waldmeister is clearly a taste of childhood: highly colored, blurred in memory, possibly nontransferable beyond a certain age.

It's got an edge of giddiness, this smell, of alarmingly loosened controls. It used to be called *Frauenbettstroh*, "maiden's bedstraw," for the springtime practice of inserting the herbs into a woman's bed pillow to waft up a suggestive scent. Roll in the hay indeed.

On the more highbrow side, you have Johann Strauss II's lively operetta *Waldmeister* and Günter Grass's *The Tin Drum*, in which the main character Oskar's first encounter with sex revolves around waldmeister-flavored fizzy soda powder; elsewhere in the book things with waldmeister get even weirder.

The dominant note in this smell is coumarin, a compound common to many plants including hay, vanilla, cassia, tonka beans, sour cherries, and dates. It's toxic enough that coumarin is banned as a food additive in the United States, although a bad headache is typically the worst after-effect. It's the kind of mild risk Germans wave away like so many tiny springtime butterflies.

Time is crucial to sweet woodruff's scent. The living plant smells like nothing; snip off a sprig and still it's odorless. If you come back hours later, though, a distinct scent will emerge. But it takes patience and some knowledge to recognize it. Waldmeister, it turns out, was the first smell I actively tried to hunt down and develop a more nuanced relationship with.

It took multiple tries to nail Maibowle, and timing was crucial. The recipe I chose instructed me to cut the fresh waldmeister sprigs and let them wilt "a while." I interpreted this as an hour, which is not nearly long enough. It called for tying together a bunch of fresh lemon balm, mint, and wilted waldmeister, cut stems up, submerging them in the wine for precisely thirty to forty-five minutes—to limit the amount of coumarin infused into the drink. Inverting the waldmeister stems was key, as a bitter, earthy-tasting liquid can seep from the cut stems into the punch. While the herbs steeped, I sliced fresh lemons and froze them to replace ice cubes, which might otherwise rob us of the delicate waldmeister flavor.[2] The resulting punch was pleasant if insipid, not perfumed at all. I resolved to try again.

With more research, I learned the sweet woodruff should actually wilt for *eight hours*. So I snipped more stems at breakfast and waited. The result by dinnertime: a tiny cluster of wilted leaves unloosed a pungent, extraordinarily "green" scent tinged with bakery sweetness. But more than anything specific, the scent evoked temporariness: a fleeting quality of sunlight and shadow, a lost bright memory of a summer's day.

I mixed up this second batch using properly wilted, aromatic waldmeister, and again effervescent wine. Again I froze the lemon slices solid prior to submerging them. The result was deeply improved. Lifting the pitcher from fridge to table left a stereophonic scent in its wake. My friends and I all plunged our noses inside the pitcher in turn; the perfume was strongest inside, like a fairy trapped in a terrarium. The time to enjoy Maibowle was exactly *now*, before the punch fizzled out, the frozen lemons melted, and the wafting scent of waldmeister escaped the interior of each of our glasses.

I didn't stop with Maibowle. I stocked up on fake waldmeister-flavored candies and syrup. I snipped and snipped at my vigorously growing sweet woodruff plants to make my own waldmeister syrup, a simple syrup infused with sweet woodruff and lemon zest. Then I made forestmaster pudding, a panna cotta in which waldmeister leaves infuse sweetened whipping cream. You add softened gelatin and crème frâiche, then refrigerate it overnight. I woke up to the usual cereal and coffee, followed improbably by waldmeister pudding. And finally, I could clearly taste the flavor. I turned to my two Waldmeistersirups, real and fake. Now I could detect their elusive similarity. I sniffed another wilting stalk. Its imprint rushed back again, detailed and immediate.

With greater familiarity, I'll now admit I find waldmeister's taste a bit one-note. But the smell is altogether different: heady, pointed, insistently green, exceedingly difficult to catch, thrilling when you do. I can easily imagine wanting to chase it again and again.

BITTER ALMONDS

What it smells like is an expert baker producing her crowning success of the holiday season. Despite its name, the scent of bitter almond oil is sweet but not cloying, with a subtle and refined nuttiness. Tilting the bottle to my wrist, the oil's scent lilts upward and blooms briefly but intensely. Almost immediately the smell vanishes, leaving only a trace of extra softness to the skin. (It's quite popular as a natural moisturizer or hair tonic.)

This is also the telltale scent suffusing many a murder scene in classic mystery novels. The Christmassy scent of bitter almonds wafting from the corpse announces, with grim irony, that this person was murdered by a dose of cyanide.

As a murder weapon cyanide is a pretty blunt instrument. But as a plant toxin, cyanide operates quite elegantly. Several edible plants—apples, peaches, apricots, lima beans, barley, sorghum, and flaxseed among them—release cyanide to kill any herbivores unwise enough to eat them. The plant stores cyanide attached to sugars as an inactive molecule inside one compartment of its cells. Inside a different compartment lies an enzyme that activates the cyanide only if needed. When an herbivore ambles into the picture and starts chewing on the plant's leaves, it crushes those two cell compartments and mixes their contents together, activating the cyanide to potentially lethal effect.

How cyanide kills a person (or a chewing animal) is similarly elegant. Cyanide prevents cells from using oxygen by interfering with the cells' biomachinery to convert food to energy. Death by cyanide is basically asphyxiation at the molecular level. This biomachinery is so basic and shared among animals, it makes cyanide an effective way to kill nearly all kinds of them.[1]

If cyanide permeates so many common foods, why doesn't it kill us when we eat those foods? The answer depends on which cyanogenic plant you're talking about. In apples and peaches, cyanide concentrates in the pits and seeds, which we throw away. In other plants, simply crushing the food and washing the mash will rinse the cyanide away. By and large cyanide is easily managed and doesn't pose a threat to the U.S. food supply.

Cassava is the only edible plant where cyanide poisoning is an ongoing issue. In Africa, people who eat too much bitter cassava and too little protein suffer from a disease known as *konzo*. If they could rinse the cassava mash more thoroughly in water, they would reduce its toxicity—but that's difficult to do when you live in drought-prone areas.[2]

It's also possible to disable the cyanide-producing gene in plants like almonds. It only took a single mutation to create bitter almonds' friendlier cousin, sweet almonds. Bitter almonds, or *Prunus amara*, taste incredibly bitter when eaten raw, and it takes just twenty nuts to kill an adult. Yet the same bitter almonds can be boiled or baked to leech away the deadly cyanide and then make marzipan or Christmas stollen with them. Sweet almonds, or *Prunus dulcis*, are equally popular in baked goods and desserts. But they lack the heavenly perfume of bitter almonds, or the frisson of mortal danger averted.[3]

Funnily enough, many poisons smell lovely if you survive long enough to appreciate them. Cyanide is one of the few poisons a victim might perceive: in how many murder mysteries does the victim catch a whiff of bitter almonds, with panic and belated knowledge rising fast behind it? Other poisonous smells register later, scenting a scene of carnage. The World War I blister agent lewisite apparently smelled intensely of geraniums. Diphosegene, another blister agent, smells of anise. Certain nerve agents are redolent of overripe apples or other rotten fruits.[4]

Natural poisons release perfumes, too, some more toxic than others. Datura is a form of nightshade with an elegant, trumpet-shaped flower. Its leaves stink but the flowers are sweet-smelling to the point of intoxication. Sniffing datura has a narcotic effect frequently harnessed as a hallucinogen for vision quests.[5] Water dropwort—also known as hemlock—smells like wild parsley or carrots. Eating even a single bite can kill and, gruesomely, leave a corpse whose face is frozen in a "sardonic grin," a rictus induced by the toxin.[6] The leaves of a belladonna or deadly nightshade smell bitter and unpleasant, but the smell of its berries has been likened to unripe tomatoes, a high-summer smell. I picture a body crumpled, in ecstasy or laziness or death, under a tangle of green tomato vines.[7]

Bitter almond oil reminds me of a basic if paradoxical fact: death can arrive wreathed in seductive perfumes; indeed, how else can it catch us unawares? A murder scene's atmosphere is more than metaphorical. Sometimes it clings to the actual air: through the cops and detectives and agitated witnesses and passersby threads the incongruous, sweet-scrumptious smell of bitter almonds.

CINNAMON

It zings. A liveliness plays over this smell like crackling electricity—at least, it does in the pricey Ceylon ground cinnamon variety I'm holding. As the scented dust shifts heavily in its Ziploc bag, it opens up different micro-tone notes. The scent almost scintillates. My grocery-brand ground cinnamon, of Vietnamese origin, lacks these shimmering top notes.

Its heat startles me. The scent of quality cinnamon resembles the slow unfolding burn of long-roasted chiles. Germans refer to anything spicy as *scharf*, or "sharp"; you crank up the spice on any food by asking for it, ungrammatically, *mit scharf*—"with sharp." Cinnamon's heat is indeed *scharf* without feeling sharp at all. It's rounded and smoldering. A penetrating sweetness pings at the tail end, smelling precisely like Red Hots. Who knew how exactly that mass-produced candy captured cinnamon's true smell?

It's Pavlovian, this smell, and as such weirdly under-interrogated. A whiff of cinnamon cues up a well-worn seasonal groove of associations: The return of colder weather. Coziness. Baking. Grandmas. Christmas. Full stop. Or is it?

The smell of cinnamon is so familiar, so reassuringly bound up in its holiday box, that we assume we know it thoroughly. But that feeling is illusory. Learning even the most basic facts about cinnamon rendered it strange and defamiliarized. Here are just a few points of interest: cinnamon is actually wood, specifically the bark of a particular tree. The Romans never ate it but burnt it copiously as incense.[1] Most of the "cinnamon" Americans eat isn't actually cinnamon at all. And this wholesome smell, now the provenance of grandmas and church bazaars, was to the ancients downright sexy.

CINNAMON STICKS ARE THE TENDER INNER BARK of a species of laurel tree, *Cinnamomum*. It flourishes in its native Sri Lanka (formerly Ceylon), but also grows in Vietnam, Indonesia, and China. You trim off the fresh tree shoots, then allow them to ferment indoors. After that you remove the twigs and outer bark, revealing the prized inner layer. You rub this bark with a smooth brass block to soften the tissues and loosen the cinnamon layer from the twig. Using a curved knife called a *kokaththa*, you slice off the cinnamon bark in a single piece. It curls into a characteristic, telescope-shaped "quill," then dries out for several days. You can grind these quills into a fine powder, extract their essential oils, or ship the quills whole as cinnamon sticks.[2]

Cinnamon's actual origins are mundane compared to myth. According to Herodotus (fifth century BCE), giant cinnamon birds in Arabia—sometimes referred to as phoenixes—used cinnamon sticks to build their nests, which they situated on sheer and inaccessible cliffs. To collect the cinnamon, Arabians would chop up oxen and toss the meat on the ground. This tempted the enormous birds to swoop down, grab the meat, and bring it back to their nests. The heavy meat hunks would snap off small twigs of cinnamon from the nests, which fell to the ground. Spice collectors supposedly dashed out to gather these twigs in the shadow of feasting giant birds.[3] Writing a century later, the "father of botany" Theophrastus described how cinnamon hunters wrapped themselves in armor to collect the plants where they purportedly grew: in deep glens protected by poisonous snakes.[4]

Cinnamon existed first as a smell, and only much later as a taste. Its history traces a similar trajectory to that of nearly all the Eastern spices: desire for its dazzling potency galvanized global trade, and later colonialism, into existence. Spices were prized first as medicines, perfumes, and incense for sacred occasions, only later on as food and drink. Spice routes girdled the planet, connecting more and more dots in an increasingly dense network, until spices poured from east to west and south to north. Their eventual abundance drove down prices and begat democratization, a trend furthered by synthetic smells and flavors in the late nineteenth century. Cinnamaldehyde, the signature note of cinnamon, was among the very first synthesized flavor molecules in 1834.[5]

Before cinnamon became common it was sumptuous, synonymous with the finest perfumes and incense. In one of the earliest written citations of the profession, the Greek poet Antiphanes describes the handiwork of a famous perfumer Peron:

I left him trying Peron's unguents,
Bent on mixing nards and cinnamons for your scent[6]

The Roman Emperor Nero famously killed his consort Poppea by kicking her in the stomach. In a pyrotechnic show of deep remorse and even steeper expense, Nero cremated Poppea's body on an enormous cinnamon pyre that burned steadily for a full day and night.[7]

Reserving costly spices for special occasions explains why so many Christmas treats are still spice-laden, with cinnamon the reigning queen among them. Mulled wine, cookies, fruitcake, gingerbread, all sweetened with molasses—all of these are direct holdovers from medieval European cookery.[8] Bakers would lay on the cinnamon thickly at the holidays and other festive events.

Smell and taste are in fact but a single composite sense, whose laboratory is the mouth and its chimney the nose.

—Jean Anthelme Brillat-Savarin

And yet the cinnamon so familiar from holiday baking is often not cinnamon at all but a related spice, cassia. You can tell the difference in several ways. First, actual cinnamon is a paler blond color. Its quills are fragile and curl in a single direction, into itself. Cinnamon's smell is milder, more nuanced and delicate. Cassia is stronger and cruder in every way: a brick-red color, with strong, stiff quills that curl in two directions (in an ornate, curly "3" shape), with a much broader and spicier smell. Cassia also contains the mild toxin coumarin, which actual cinnamon does not. Some aficionados favor true cinnamon for baking, cassia for meats and savory dishes.[9]

Perhaps the biggest surprise of cinnamon is how sexy the ancients found it. Along with cloves, nutmeg. and especially ginger (see the ginger section in chapter 2), cinnamon was considered surefire for stoking lust. A harlot in the biblical book of Proverbs used cinnamon to lure a young man to bed "as a bird rushes into a snare." Similarly, Song of Songs reeks deliriously of cinnamon, honey, spikenard (a fragrant flower), frankincense, myrrh, saffron, cloves—all the sexiest spices.

The Roman playwright Plautus wrote a comedy called *Casina* (*Cinnamon*) in which the title character is a clever minx whose sexual pull incites all the action in the play. At one point a drunken character moans to Casina: "My beauty of Bacchus! . . . Compared with you, every other essence is bilge water! You are my myrrh, my cinnamon, my ointment of roses, my saffron, my cassia, my rarest of perfumes! Where you are poured is where I want to be buried." In *The Golden Ass* by Apuleius, the main character is driven mad with lust by a slave girl's jiggling buttocks and her cinnamon-hot breath. In Virgil's *Aeneid,* Cupid visits Psyche invisibly at night, and yet she recognizes his presence by his "cinnamon-fragrant curls." St. Augustine reviled his youth misspent "walking the streets of Babylon, in whose filth I rolled as if in cinnamon and precious ointments." Downing an amphora of hippocras—wine steeped with cinnamon—readied the Romans for a night of debauchery as well as debate.[10]

Constantine the African was a medieval Benedictine monk who translated Arabic philosophical and scientific texts into Latin, reintroducing a standard of rigorous inquiry lost to Europe after the Romans fell. Despite broad contributions to European intellectual life, Constantine is now perhaps best known for authoring the foremost sex manual of the Middle Ages, *De coitu* (*On Sexual Intercourse*). In it Constantine recommended cinnamon in all kinds of electuaries, spice-recipes with a purported medicinal effect. One might boost a flagging sex drive with a mixture of galangal, cinnamon, cloves, long pepper, arugula, and carrot; "the best there is," declared Constantine, kissing his own (apparently un-celibate) fingers.[11] In *The Perfumed Garden*, a fifteenth-century Arab sex manual, author Sheikh Mohammad al-Nefzaoui advises chewing up a mixture of cinnamon, ginger, "cubebs," and "pyrether" (two North African spices), rubbing this same mixture copiously on one's penis and then driving a woman wild with it.[12] As late as the eighteenth century, English newlyweds still drank a "cosset" before proceeding to the marriage bed: a mixture of wine, milk, egg yolk, sugar, cinnamon, and nutmeg.[13]

So here is cinnamon: the spice heralding cold-weather coziness, but also ample quilt-diving and bed-romping. You might say it's the spice that warms *all* your senses. The spice that fills you with indulgent ideas, that fills you itself with its spreading warmth. Grandma's wink when she delivers those cookies might be more knowing than you realized.

HOT CHOCOLATE

The scent starts off molten, tarry, slightly dank at its base. There's a pleasing astringency, veined with iron notes. The smell fills your nostrils substantially in the way a dense paste of pulverized beans ought to.

But oh, the liftoff! Curling in zephyrs above the mug, hot chocolate creates above itself a dome of delectable, shifting scents: floral, sweet, tanged with citrus, a slight chili heat. That cloud of sweetness can hit the nose as crystalline and brittle, like grains of brown sugar. Sniff again, though, and it may read sweet in a liquid, amber, malt-syrup way. You could close your eyes and focus solely on sniffing how much milk the drink contains. You can play a nearly endless game of blink and sniff, blink and sniff. Each time until your nose exhausts, many different scents lift magically from its liquid surface.

Chocolate's smells are super-dimensional. Like coffee, tea, beer, and wine, chocolate has tons of aromatic compounds—specifically six hundred, compared to two hundred in wine. Only twenty-five of those compounds determine chocolate's olfactory profile, together signaling a recognizable "chocolate" smell and flavor. But those twenty-five defining compounds are a motley bunch and not all individually pleasant: many of the same molecules figure prominently in the smells of human sweat, cabbage,

Compare similar smells.

Take out every bottle of vinegar you own and pour some of each into its own small bowl. Above their common base note of vinegar, can you perceive other differences between their smells? Apple cider vinegar is spiky and alcohol-forward. Sherry vinegar is mellow and woody. Even white wine and balsamic vinegars, both made from grapes, smell markedly different.

Try this exercise with other pantry collections you've amassed, like cooking oils—avocado, almond, coconut, sesame, olive—or different kinds of salt. Or buy a variety of fresh citrus: limes, lemons, grapefruits, kumquats. If you have trouble observing different smells, try warming each ingredient gently over the stovetop to release the smells more.

Bonus:

Repeat one of these smell-comparisons with a friend's help. Ask her to mix up the dishes so you can sniff each smell with your eyes closed. Can you identify which smell is which?

and beef. When two chemists managed, in a 2019 study, to nail the compounds and concentrations required to synthesize the chocolate aroma, it was a triumph for chocolate manufacturers everywhere.[1]

Our brains can't handle more than four competing smells at once before we're befuddled. Even experts have difficulty differentiating more than three smell compounds simultaneously.[2] It's this enjoyable befuddlement that explains chocoholism at the gourmand level. We love trying, and only partially succeeding, to tease apart chocolate's many smells. Chocolate is not so much *one* smell as a detailed world of smells you can enter and explore at will. Every whiff yields a fresh crackle of nuances.

CHOCOLATE WAS FIRST CONSUMED AS A HOT, aromatic drink. Hot chocolate as drunk by early Mesoamericans was very different from our modern recipe. Mesoamericans would roast and grind the cocoa beans into a paste, mix this with cornmeal, then flavor it with spices like chili pepper, vanilla, annatto paste, and various flowers. They'd mix this paste into hot water, then pour the liquid from a great height into its container—a gourd or clay pot—creating a characteristic foam on top of the drink.[3] In this form, hot chocolate was more of a savory liquid meal than a sweet confection. It was portable—when dried, chocolate paste lasted a long time and could be reconstituted into a tasty drink anywhere simply by stirring it into hot water. Mayans and Aztecs often used hot-chocolate paste and cocoa beans as a universally accepted traveling currency.[4]

To the Aztecs, the smell of hot chocolate was a heady brew of bitterness, liquid gold, adrenaline, and liveliness. Coaxed forth by heat, a constantly shifting array of smells drifted up from chocolate's foamy surface. Its elixir status is confirmed by evidence that the Aztecs offered sacrificial victims chocolate mixed with blood-stained water in their final moments of life.[5]

Among Europeans colonizing the New World, women began drinking "Indian hot chocolate" before men did. In one conquistador's account, Spanish women persuaded their maids to sneak hot chocolates to them during long Catholic Masses, enraging the priests as the heady scent threaded its way through the pews.[6] Hot chocolate became more broadly popular among sixteenth- and seventeenth-century Spanish Catholic colonists during fasting periods. The rules allowed penitents to consume drinks during the daytime, but no food—and hot chocolate was the perfect loophole snack, tastier and more sustaining than eggnog, soup, or gruel.[7]

In the late sixteenth century, Spanish Jesuit and Dominican clergy shared bits of chocolate with their brethren living in other countries. Monk to monk, priest to priest, this is how chocolate first infiltrated Europe.[8] Initially Europeans changed the Aztec drink very little: instead of pouring the liquid from a height to aerate it, Europeans used a tiny whisk called a *molinillo* to create foam and stir up the drink's scent.[9] Eventually hot chocolate's recipe morphed, which naturally changed how the drink smelled. Starting in the early eighteenth century, Europeans added hot milk, sugar, eggs, and even custard to hot chocolate.[10] They drank their chocolate full-fat, without skimming off any of its abundant cocoa butter—not unlike the modern keto fad of drinking coffee with a stick of butter melted into it.

Chocolate's preferred natural state is molten; it actually takes a lot of technical doing to turn it into a solid. Chocolate only solidified into a bar shape after the invention of "dutching" in 1828. This process separated the cocoa butter fats out from chocolate liquor, creating a less fatty paste that manufacturers sweetened and sold as hot chocolate mix. A few decades later, manufacturers ground up this less fatty chocolate paste even more finely and combined it with sugar and milk powder. They added some cocoa butter fats back in, then "conched" the resulting chocolate mix for hours—a mechanized form of extreme grinding and mixing to turn the chocolate mix from grainy to silky smooth. This entire process warmed the chocolate to 60–70 degrees Celsius, so the final step was "tempering," rapidly cooling the chocolate to make the cocoa butter form crystals to stabilize its structure into a solid.[11] And thus the chocolate bar was born.

Chocolate always wants to undo this laborious process and become liquid again. Much like humans, chocolate sweats at temperatures above 80 degrees Fahrenheit (25° C) and melts soon after that. Maybe that's why its smell can be so sexy: chocolate is responsive and pliable. It melts at your slightest touch.

Like all colonial goods, the story of chocolate can get really, really dark. West African countries, chiefly Ghana and the Ivory Coast, supply more than 70 percent of the world's cocoa beans[12] and both countries are still plagued with illegal child labor that's tantamount to slavery. Many workers on cocoa plantations have never tasted the final product, and too many of these workers are also children.

Fair-trade and single-origin certifications for chocolate are bringing badly needed transparency to chocolate production, which helps reduce instances of child labor and deforestation while elevating the product's quality.[13] Ethical chocolate producers want to transform this historically exploitative supply chain similar to the way coffee production has already been radically improved. So far, they seem to be succeeding.

Sniffing my mug of hot chocolate again, I think less about mugs past and more deliriously large-scale: to entire city blocks scented with chocolate. I live in Chicago where the Blommer's Chocolate Factory pumps out chocolate smell to the West Loop so regularly, there was for a time a popular mobile app to help fans pinpoint its scent on a given day. My first encounter with Blommer's brought back another nested memory. Once years ago, we followed a secret bike path in Berlin with our friends Heidi and Thomas. It wandered due north along the path of the former Wall, the demilitarized zone now planted over with trees that were mere saplings when I biked past them. A chocolate smell wafted along our way seemingly for the entire trip, threading through the slender trees. I recall the breezes shifting just enough that the smell waxed and waned but then waxed anew, and our noses never exhausted on the scent. We spun along in happy silence, in dappled sunshine, following an aerial ribbon of chocolate. I return to Berlin often enough now that I could try to re-create this bike trip. But the memory is too beautiful and fragile to interrogate or even to repeat.

3 SAVORY

Ranging from classic favorites to lesser-known smells, these sections deal with finding satisfactions wherever you can and learning to love the off-putting and unfamiliar.

BACON

DURIAN

STINKY CHEESE

ASAFOETIDA

TOBACCO

BACON

Raw bacon is unlovely, gelid, and fatty. It smells only faintly of salt. You put it limply into a skillet, crank on the burner, and wait. As cooking theater goes, bacon offers a very cold open.

Bacon's smells are kinetic; they're spun dizzily into motion only with the application of heat. As the skillet warms, a succession of smells layer over themselves in an ever-deepening bower. The smell holds the promise of good pork and fat's full, rounded savor. It's sweet, syrupy, crystalline. It pings satisfyingly with brine. You might catch a whiff of smoke edging the smell delicately; one sniffs for that note constantly, checking for imminent burn.

It's multisensory, this smell, impossible to extricate from the action inside the pan. You're both watching and hearing a smell develop—a smell with a tangible nimbus above the pan, microdroplets of spitting grease. As the skillet warms, the bacon's edges lift and curl as if alive. Then the strips sweat profusely, releasing an amber lake of hot grease. Then they shrink as they brown. And all the while, smell after smell arcs upward like volatile bursts of fireworks. It builds to a thrilling crescendo at the peak of cooking and casts a hazy afterglow that lingers throughout the house after the burners are switched off.

What you're smelling with bacon is a particularly fine example of the Maillard reaction, named for early twentieth-century chemist Louis-Camille Maillard. He observed a chemical reaction between amino acids and sugars when food cooks. As food heats up, sugars within the food break down, react with amino acids, and release many delectable-smelling compounds, mostly hydrocarbons and aldehydes. The Maillard reaction explains the alluring smell of any food that browns as it cooks, grills, or bakes.[1]

Bacon's smell is unique in that it's been brined or cured in salt, which makes it more nitrate-rich than other cuts of pork. The nitrates also break down when cooked, releasing an additional wave of nitrogen-related molecules—specifically pyradines and pyrazines. The white fat strips contribute a third group of aromatic compounds, with the smoke's drifting scent as a fourth. All these compounds cook at different rates, each releasing their smells in turn. The result is a layered, multidimensional scent that uncannily reflects the temporal stages of bacon cooking.

Exploring the outer limits of bacon's merchandising appeal is something of an American pastime. A lightning round of bacon-themed products currently on sale: Soap. Lip balm. Air freshener. Waxed floss. Band-Aids. Temporary tattoos. Such abundance suggests you can move any kind of merch by adding bacon smell, but the Wake N Bacon clock on TV's *Shark Tank* would prove a strong corrective to that idea. A clock that would wake you up to the smell of bacon frying, this is widely considered the worst product pitch in the show's history.[2]

Bacon defenders will be chuffed to learn that bacon's superior smell is confirmed by science. One 2004 academic compared the smells of bacon to those of pork loin, a less fatty and uncured cut. Their findings confirm that nitrates widen the amplitude of bacon's smell.[3] What's more, the bacon smell's appeal is also confirmed by math. Two different studies applied big-data-crunching algorithms to recipe websites to analyze the effect of adding bacon to a recipe. Did it tend to improve the dish's popularity versus the non-bacon-enriched version? The answer is yes, but only marginally. Adding bacon lifted a recipe's popularity rating from 4.13 stars to 4.26 stars, 15 percent of a standard deviation. Bacon doesn't make *every* dish taste better but across the boards it does produce a reliable statistical lift. Bacon mic drop![4]

DURIAN

It's commanding, even muscular, this smell. It hits the nose like ramming into a solid wall. Like Charlie Brown's friend Pigpen, durian brings its own dense cloud everywhere it goes. Stick the fruit inside any container, and its smell suffuses the space instantly and completely. With other smells, you have to sniff and sniff to find wisps threading weakly through the air. Not durian. Its smell finds you.

Rotten eggs. Roasting onions. Soup broth. Sweet lychee. Hot garbage in summer. A boiling pot of ammonia. Ripe mango. Well-spiced cabbage. Even, as my husband noted with surprise, asafoetida (see section below).[1] All of these well-publicized descriptors do fit, but sniffing durian resembles nothing so much as that old joke about three blindfolded men slapping around different parts of the elephant and describing the animal based on what each of them feels. That is, their separate observations are true but also incomplete.

Durian is extremely polyphonic. Remember the scary music used in the Stanley Kubrick film *2001: A Space Odyssey*? Its composer György Ligeti reveled in the concept of "micropolyphony." That music is monolithic yet detailed, with separate filaments shining darkly within a huge wall of sound. In its bigness and complexity it's also terrifying, like a glimpse into the mind of God. Durian smells like that.

Or, to make a visual comparison, it also reminded me of the philosopher Ludwig Wittgenstein's impossible color reddish-green. Ordinarily we cannot see these two shades distinctly when superimposed on each other—they should flatten into a muddy brown. Durian's two dominant smells—the stink and the sweet—should similarly cancel each other out. But instead they vigorously circle each other, each vibrant and undampened, an impossibility made real.

When you taste durian, the myriad weirdnesses merge and resolve, but only imperfectly. A warm, honeyed tropical sweetness fills the foreground, melting instantly over the tongue. The fruit's texture is languid, soft, faintly buttery—a fattiness unusual in fruits that makes both the smell and taste linger intriguingly. It crackles with a slight spiciness which, when consumed in large quantities, can numb your tongue slightly. Durian fans find this sensation exhilarating in the same way eating crazy-hot chilis can be. The fruit's sweetness would be almost cloying, if not tempered by its unholy stink.

Durian maxes out your senses. The fruit looks straight out of Pokémon: a large green pod covered with uncannily perfect, geometric spikes. You cannot cut into a durian without wearing oven mitts to brace against the spikes; baseball mitts also work. Yet when the knife cracks the fruit's exterior, it cleaves open lobes of soft flesh, pale yellow to deep orange depending on how ripe the fruit is. You tenderly remove each lobe and remove the hard orange seed at its center. The lobes are breast-like, round and heavy with juice, reeking at full blast. My husband did the carving and described it as "literally the kinkiest thing ever, disgusting and hot and earthy and dewy-sweet."

Durian's smell polarizes. You either love it or hate it, or you don't know it at all. I was in the third category, curious to form my own impression. What I knew going into this encounter was that the stink is notorious but beloved by aficionados. Native to Thailand and Malaysia, durian is prized in season (June) and suitably expensive. With an extraordinarily high natural sugar content, durian is popular in desserts of all kinds: candies, moon cakes, sticky rice, and especially ice cream. (But it's never added to alcoholic beverages: durian strongly inhibits the enzyme that the liver relies on to metabolize alcohol.[2] There's some truth to the old-wives' saying that eating durian while drinking booze can kill you.) Many Southeast Asian countries have banned the fruit from public-transit systems, airports, and offices. It's a room-clearer pretty much anywhere it's consumed.

You have to wonder: why does durian stink so much? Like any other fruit plant, the most obvious reason is to attract animals—elephants, rhinoceros, tigers, civets—to eat it and spread its seeds. These animals roam through very scent-heavy jungles; perhaps durian needs to fight above other fragrances to get noticed? A 2016 analysis of durian's odor compounds found it has an extremely low odor detection threshold. That

Collect new smells.

Survey your house for things you hadn't thought to smell before. Start in your kitchen.

The ordinary world abounds with smell-surprises. My white sugar bowl smells so convincingly like salt, I'm tempted to lick it to make sure it's really sweet. Citric acid—a natural household cleaner—smells deep-roasted and almost nutty, not like citrus fruits at all. The scent of wool hints at its secret waxiness. Your child's stuffed animal smells like her hair.

If you have a mortar and pestle, grind up whole spices into powders and smell those. Coriander seeds, for instance, smell keenly of citrus with grassy green notes. Make a list of your smell-surprises.

is, for durian's most dominant smell notes, you don't need very many scent molecules in the air before average people can start to detect it.

Vanillin, the molecule most responsible for vanilla's scent, also has one of the lowest odor detection thresholds going. At 0.1 or 0.2 micrograms per cubic meter, one or two oil tankers full of vanilla extract could scent the entire planet lightly with vanillin.[3] (For more on this smell, see the vanilla section in chapter 2.) By contrast, durian's predominant smell molecules have odor detection thresholds as low as 0.00076 (that's for the molecule 3-methyl-2-butene-1-thiol, imparting a "skunky" scent) and 0.0080 (ethyl (2S)-2-methylbutanoate, giving durian a "fruity" note).[4]

Smelling [the perfume *Feu d'Issey*] is like pressing the play button on a frantic videoclip of unconnected objects that fly past one's nose at warp speed: fresh baguette, lime peel, clean wet linen, shower soap, hot stone, salty skin, even a fleeting touch of vitamin B pills, and no doubt a few other UFOs. . . . Whoever did this has that rarest of qualities in perfumery, a sense of humor. . . . Whether you wear it or not, it should be in your collection as a reminder that perfume is, among other things, the most portable form of intelligence.

—Luca Turin, *Perfumes: The Guide* (2008)

A strong smell, unavoidable and lingering, can bulldoze over you in an unpleasant way. Durian's scent thoroughly busted my previous sense of the category of "fruit" and the possible smells a fruit might contain. When I let these associations go and stopped trying to resolve my cognitive dissonance, durian got easier to smell. Similarly, accepting that this flavor is prized by another culture—literally billions of Asians cannot be wrong—made me stick out the stink and search for what's transporting within it.

Eating durian also changed my perception of its reek. Like stinky cheese, the smells change when inhaled retro-nasally (inside the mouth). Allied with its tastes and textures, durian fused its disparate strains in an ever-shifting flavor that manages to stay both coy and emphatic. It's always giving you more facets of itself to observe.

It's this prismatic quality of durian's smell and flavor that might damn it to the un-initiated. In 1998, Pam Dalton of the Monell Chemical Senses Center was tasked by the U.S. Department of Defense with producing a universal stink bomb. The resulting compound, called Stench Soup, is a solid contender. But creating a truly universal stink bomb, guaranteed to repulse everyone, is trickier than it sounds. Even the most obvious repellent smells—sewage, rotting trash, dead bodies—are common-enough smells in certain parts of the world that people can get habituated to them. The ideal recipe for a stink bomb combines gross and delicious smells in an unexpected amalgam. The composite smell should be simple enough that we can discern the different notes within it, but those notes should feel discordant when combined. Stink is a matter of uncomfortable proximity, a sense of shitting where you eat.[5] If you're not prepared for it, durian's smell can check all those boxes.

We made our raw durian fruit into ice cream and found its smells and flavor transformed yet again. Far from subduing the reek into creaminess, durian ice cream foregrounds its unnatural perfume. As chilled ice cream melts on the tongue, the flavor emerges warm, animated, and rich on the palate. Like a fine fetid cheese, you savor the complex odor secure in the knowledge that it won't hurt you. Much as the British naturalist Alfred Russel Wallace described durian, the range of flavor associations is dizzying. Wallace wrote of durian in 1856: "A rich custard, highly flavoured with almonds . . . [with] occasional wafts of flavour that call to mind cream-cheese, onion-sauce, sherry-wine, and other incongruous dishes. . . . the more you eat of it the less you feel inclined to stop."[6]

It's unlike anything I've ever tasted. Durian's stink shimmers with liveness.

STINKY CHEESE

Feet! FEET! From inside a round wooden box the smell rises in waves like a rolling miasma. It seems to pulse with intensity. It waggles its very unwashed toes. Cradling the box in my hand, with its contents encased in fluted wax paper, one feels the perversion of an overly close excavation of something intimate, noxious, unstable.

The outer rind of my cheese is orange and mottled in a lacy pattern. Sniffing an inch or two from the rind produces a concentrated, potent, highly interesting little cloud. It feels like meeting an alien creature livid with unfamiliar energy, which breathes nervously as it rests in your palm. That equilibrium is fine so long as it remains undisturbed. Sniff the cheese from any closer, however, and your nostrils shut down as if seared by chemical blast.

I am sniffing—but not yet eating—L'Ami du Chambertin, a cow's-milk cheese made in the same fashion as Époisses, deemed the world's stinkiest cheese in a 2004 study[1] using electronic noses.

Stinky cheeses acquire their smells in three different ways. Blue cheeses like Roquefort and Stilton are soft, open-structured cheeses made from milk mixed with blue mold spores called *Penicillium roqueforti*. The cheese's innards (called by experts the "paste") are pierced to allow air to enter; the mold spores bloom and produce the blue veins characterizing these cheeses. Bloomy or brie-style cheeses like Camembert are similarly mixed with spores of *Penicillium camemberti*, which produces the fluffy white outer rind. Much like its cousin *roqueforti*, this fungus softens the inside cheese but also makes it reek alarmingly of ammonia. In both cases, fungi and other microbes transform the cheese from the inside out and produce a magical sensory divergence: supreme butteriness, with innumerable subtler notes, contrasted with a blast of unholy funk.

The stinkiest cheeses of all are those that have their rinds washed in brine or liquor before aging. L'Ami du Chambertin and Époisses are both washed in *marc de Bourgogne*, a type of local brandy made from the leftover dregs of winemaking. This wash encourages the growth of *Brevibacterium linens* and other types of coryneform bacteria, which produces a vivid orange rind along with an extraordinary stink. These cheeses smell like dirty feet because it's exactly the same bacterium at work. Washed-rind cheeses often become so soft they're served in a wooden box and can be eaten with a spoon like pudding.[2] Other famous washed-rind cheeses—and they all tend toward notoriety—include Limburger and Vieux Bologne (washed in beer), Stinking Bishop (washed in perry, a drink made with fermented pear juice), Munster (washed in brine), and Hanvi (washed in *marc de Gewürztraminer*, a brandy made from already-pressed Gewürztraminer grapes).[3]

Why on earth would you eat something that stinks so keenly? You wouldn't unless pressed—but learning to overcome that initial resistance, to train the palate, is an exhilarating experience for many. And learning how stinky cheeses get made can also help coax reluctant eaters along.

TURNING MILK INTO CHEESE BEGAN AS A PRACTICAL MATTER. When milk is rendered solid, it becomes portable, lasts much longer, and tastes more interesting. Bronwen and Francis Percival's book *Reinventing the Wheel* gives the best explanation of cheesemaking 101. You take milk from a cow, goat, sheep, or some other mammal. You introduce an enzyme (rennet) which induces the proteins to clot, trapping fat solids as well. You then add starter cultures to kick-start the fermentation process. Next you separate out the moisture (whey) from the milk curds. Lastly you store the solid curds under certain conditions so they can age and change.[4] All the variations in cheeses can be explained by varying one of these four factors: the type of milk you start with (and whether you pasteurize it first); when you choose to start and stop the fermentation process; when you separate the curds from the whey; and what you do to the curds before storing it in a particular, well-chosen place. Much like a doctor's role in a healthy pregnancy, cheesemakers are often just well-informed observers who watch a natural process unfold and only occasionally intervene in it.

Stinky cheese is a niche enthusiasm with a very long provenance. In the Sumerian-Akkadian languages of that ancient civilization, there's one word for the broad category of cheese, *ga-har* or *eqidum,* but another reserved for stinky cheese, *nagahu.* That word also served nicely as an insult.[5]

A medieval European biographer, Notker the Stammerer, records how Emperor Charlemagne was introduced to blue cheese during a dinner visit to a bishop that happened to fall on a Friday. Forbidden as a Catholic from eating meat that day, the bishop served up his best cheese which Charlemagne delicately picked at, avoiding its unfamiliar and putrid-smelling blue veins. The bishop pressed him to try that part as the most delicious bit. Charlemagne assented, with rapidly increasing enthusiasm. By the end of the visit, the emperor was so convinced he ordered a regular supply of blue cheese to be sent to his capital.[6] In the early eighteenth century the writer Alexander Pope thought Stilton the finest cheese a country mouse could dream of. His fellow writer and contemporary Daniel Defoe ate Stilton gustily "with the mites or maggots around it so thick that they bring a spoon with it for you to eat the mites with."[7]

Maggot-riddled cheeses are an extreme outlier of the stinky-cheese enthusiasm, but very much a real and modern Thing. In a Sardinian cheese called *casu marzu,* cheesemakers deliberately introduce *Piophila casei*, or cheese flies, to lay eggs inside the cheese. The larvae eat the fats in the cheese, which produces an advanced stage of decomposition and a super-silky, if pungent, cheese. Debate rages over how to pinpoint the cheese's fine line between optimal flavor and actual rot. Some eaters advise storing the cheese in a paper bag, which deprives the maggots of oxygen. They pelt themselves against the bag, trying desperately to escape. Ideally one should eat the cheese at the moment the pelting sound stops.[8]

What exactly do we fear in stinky cheese? This, too, has been studied by science and awarded a 2017 Ig Nobel award, a dubious distinction for scientific findings with a ridiculous twist. Turns out cheese offers a perfect object lesson in the neurology of

For I will give you a perfume,
That the very Venuses and Cupids gave my girl;
When you smell it you will beg the gods,
Fabullus, to make you all nose.

—Catullus

disgust. Remember: as a sense, smell is designed to detect potential threats and is therefore judgy. Smell a super-ripe cheese for the first time, and your mind will race with ill-informed notions: Is this cheese incredibly old, or bacteria-riddled in a way that could be worrisome? What good could possibly come of ignoring the nose's rank evidence and *eating* this mess?

Disgust moves speedily and invisibly in the mind, like microbes themselves. The 2017 study separated its French participants into avid cheese-lovers and cheese-haters. They served up cheeses to participants while performing fMRI scans of their brains. The reward centers associated with perceiving food actively shut down in the cheese-haters' brains; at a precognitive level, their brains refused even to perceive the cheese *as* food. The brain's insular cortex is also involved in disgust, an area also associated with self-awareness. If you are anything but an avowed stinky-cheese lover, your insular cortex will conjure a mental image of yourself, hunkering down to eat a plate of iffy-smelling cheese. Seen from this skeptical view, you look and feel like a ripe old patsy. Your will rises up rebelliously. *La confiance, c'est un jeux de dupes!* Trust is a fool's game![9]

But this line of thinking is redolent, more than anything, of not knowing the facts. Stinky cheeses don't necessarily push the limits of milk spoilage more than sweet-smelling cheeses. They're also not always extremely old, although washed-rind cheeses do have a vigorous mat of bacterial and fungal biofilm growing all over their exteriors. In other words, stinky cheeses don't embody the dangerous microbial brinkmanship that we might secretly fear they do. What they do offer is access to a lively, multivalent, and changing amplitude of flavors.

In the process of its creation, cheese quietly absorbs many inputs. The mammal eats its favorite local foods, and those terroir tastes enter its milk. Cheese reflects the stages of its fermentation and curdling, what was done to it when. And most importantly, cheese absorbs the atmosphere in which it aged. It sits and changes inside cavernous walls that may be steeped in centuries of friendly microbes, breathing in and out. Cheese tastes like time, experience, and locality, because it embodies all of those things. Each bite of the resulting product is only a snapshot, a candid from the hectic photo roll of life.

Smelling a stinky cheese with your nose is only one data point. But the combination of tasting it with retro-nasal smelling produces a delectable alchemy of contrasts.[10] When I take a bite of my L'Ami du Chambertin, it's a sharp, crackling vivid explosion on the tongue that melts and scintillates into an entirely different effect, like those showstopper fireworks at the Fourth of July finale. It's garlicky, eggy, almost heated in pungency, then meltingly soft and rich as the tastes sieve slowly along the tongue. So many flavors compete for space, you cannot observe them all in a single bite. So I scoop and repeat, scoop and repeat. The cheese round gets whittled to a tiny, smeared nub. The vaporous cloud of stink seems to vanish; there's only digging into this strange, unfolding heart of flavors. It's a temporary and delectable fever.

Only hours later, when my fridge exhales the full noxiousness of the scent again, do I realize how momentarily insane I was—and how delicious insanity can taste.

ASAFOETIDA

It punches outward, this smell, clearing its own space. It seems brawny, unapologetic, outsized. But after the initial thwack, the smell mellows and deepens, suggesting a landscape: a field of ripe onions sweating lightly in a heatwave. As the finer details of the smell open, the field alters. You picture fat onions drowsing between springy garlic tops, ramps, scallions. A glorious field of alliums of every description. It takes less than a minute to start feeling hungry.

Asafoetida is an incredibly stinky plant resin, a spice often substituted for onions and garlic. Nicknamed "devil's dung" in numerous languages, the name asafoetida comes from a Latinized version of the Persian word *aza,* meaning "resin," and *foetid*, Latin for "smelly" or "fetid." Modern Indians call the spice *hing*, which derives from Sanskrit *han*, "to kill."[1]

It comes from a plant that's Seussically strange in appearance. Imagine a phosphorescent lime-green Queen Anne's lace that grows to the monstrous size of a tree.

Asafoetida plants thread the desert airs of eastern Iran and Afghanistan with their reek. To harvest the spice, you tap the plant's root in springtime and let the resin ooze out. It dries into marble-hard balls that must be triple shrink-wrapped to contain their smell.

Asafoetida is a stink that can be spun into culinary gold. Smelled in its raw state, asafoetida is almost intolerably potent. Heated in oil, though, it blooms with an umami flavor perfect for vegetarian dishes. (Heating it also breaks down the sulfides that smell the worst.)

In the Middle East, asafoetida first entered written history in eighth century BCE where it was listed in the garden inventories of Babylonian King Marduk-apla-iddina II as well as a catalogue of medicinal plants in King Ashurbanipal's library in Nineveh (near modern Mosul, Iraq). During the Roman era, the plant was traded across the known civilized world: from Italy and Libya clear across the southern half of Asia. Now asafoetida is a staple spice only in South Indian, Persian, and Afghan cuisines, and largely unknown to Westerners. (A notable exception: asafoetida found its way into the recipe for Worcestershire sauce, a result of the British colonizing India.)

Classical Romans began using asafoetida after Alexander the Great returned from an expedition into northeastern Persia. They raved about a plant they'd found that was nearly identical to silphium, a fabled spice from Cyrene in North Africa. Writing in the first century CE, the eminent Greek physician Dioscorides compared the Iranian plant unfavorably to silphium. He found it "weaker in power . . . [with] a nastier smell." Nevertheless, asafoetida could be substituted for silphium in cooking—which was fortunate because a few decades later Cyrenean silphum went extinct. Romans stored asafoetida raw in jars with pine nuts. They would remove the flavored nuts and cook with these, getting a more indirect version of asafoetida's pungent flavor.

Cooks and physicians alike appreciated this plant's varied uses. In his treatise *De Materia Medica,* Dioscorides considered asafoetida almost a cure-all. In his estimation, it was good for goiters, baldness, toothache, and lung diseases from pleurisy to bronchitis. You could apply it to scorpion bites, and (if one dared) could even "cast off horseleeches that stick to the throat" when "gargled with vinegar." Modern physicians might quibble with those claims but not with these: asafoetida aids digestion and reduces flatulence, which makes its modern use in Indian *dal* (lentil) dishes both delicious and functional.

Anyone with a sensitivity to eating onions or garlic should consider asafoetida a go-to substitute. That's how I started cooking with it. I marched into Chicago's Spice House looking for exotic stinks, only to learn that asafoetida would fit the bill *and* fix my husband's allium problem. That night we made a chili that called for a boatload of onions and garlic. We used asafoetida instead: tasted great, happy tummies afterward.

Asafoetida spread throughout Europe and was used in cooking there through the early Middle Ages. The French enjoyed seasoning barbecued mutton with it. While it eventually faded in the West, asafoetida flourished in cuisines outside the Western world. With the rise of Islam in the seventh century, the Abbasid Empire gained control of Persian territories where asafoetida grew. The Abbasid caliphs oversaw a cosmopolitan court culture yielding many Arab cookbooks full of recipes for asafoetida-rich stews.

"Devil's dung" traveled to India during the rise of the Mughal Empire in the 1600s. In Agra and Delhi, another delightful (if factually iffy) medical use for it emerged: to improve the voice. Court singers would rise at dawn, eat a spoonful of asafoetida with butter, then go to the riverbanks to practice singing at sunrise. Asafoetida further gained favor in India among strict practitioners of Jainism. All Jainists are vegetarians, but a subset of these do not eat root vegetables like potatoes, onions, and garlic. These foods are considered *ananthkay*: one body containing infinite lives. Potatoes grow eyes, scallions can sprout anew if soaked in a glass of water. Harvesting root vegetables means uprooting and killing the entire plant (not to mention any micro-organisms clinging to the root) and thus ending its cycle of life. Asafoetida provides a dash of onion, without destroying any reincarnating plants. Asafoetida gives rogen ghosh and many dals its distinctive taste; it's a flavor you can't approximate merely with onions.[2]

My bottle of asafoetida powder is tiny, plastic, highlighter yellow, and curved like a woman's figure. Indeed the bottle features a picture of the goddess Vandevi. Crack open the top and you'll find a powder, the same highlighter yellow in color, that packs a wallop with just a few grains. It's so potent in its pure state, asafoetida is usually mixed with gum resin, rice flour, and turmeric to make it easier to store and to temper the stink.

My seven-year-old son shares the tiny yellow bottle with his friend in a flood of enthusiasm. Smash cut to two children with yellow-stained hands, giggling and haring around the room as it fills with locker-room smell. It blooms and blooms, an exotic herb I'd never heard of with a vast history unrecognized in the West. Yet when you eat it, it tastes like the most humble and familiar of foods, onions, and garlic. A stinky and intriguing doppelganger.

TOBACCO

The first tobacco I sniff is pure Virginia, smelling like the best kind of barnyard. It speaks of hot weather and cut grass. Its sweet syrupy-ness suggests raisins, dates, figs. This tobacco happens to be cut in flakes, delicate slivers the size of chewing-gum sticks. It's reddish-brown in color, striated like particle board. Blond streaks running through it show where the sugars have concentrated. Those sugars give Virginia an enjoyable sweetness to smoke but also make it burn extra-hot. Smoking pure Virginia is tricky for the inexperienced pipe smoker. Its scent is balanced and affable, qualities it retains when you burn it. It's not hard to see why Virginia is considered a classic.

Next we open the burley. This tobacco is darker in color and smells denser, mustier, less sweet. A whiff of forest floor clings to it, suggesting vegetal decay or sod. Burley seems more serious than Virginia, less appeasing. When you light it, burley tastes identical to smoking a cigarette. It's not a tobacco you'd choose for flavor—but then again, burley runs high in nicotine and gives tobacco blends a stimulant punch. More than other tobacco types, burley delivers the drug; it doesn't need to woo you with flavor.

Finally we crack open the Latakia. Its smell is more syrupy but not sweet, like molasses. Many top notes play over its surface: a citrusy acid, baking spices, a hint of smoke, woodsap, mushrooms. Latakia feels like a dark horse, a niche taste. It's the same leaf as nearly all smokable tobaccos, *Nicotiana tabacum*, but cultivated in a different region— traditionally Syria, now Cyprus. Latakia is the same tobacco plant responding to another climate's growing conditions.[1]

Across the table from me sits our friend Dave, the owner of these tobaccos and many others besides. He tells me Virginia, burley, and Latakia are a kind of primary colors of pipe tobacco. At least one of these appears in nearly every tobacco blend he's tried.

Dave brings the zeal of a recent convert to this new hobby, pipe smoking. He's acquired it under some duress. He's a painter toughing out a weird situation: on March 15, 2020, he was supposed to move to Berlin to start a dream job as professor of painting at an arts university. Lockdown canceled his departing flight, and the pandemic has stranded him and his wife, Jinn, in Chicago.

Dave is a laconic guy, with a dense black beard and a strong aura of reserve. He's not prone to complaint, but how he spends his days now gives him ample reason to. He and Jinn have been waiting to leave for six months. Now in the seventh month, their paperwork looks auspicious enough to book another flight.

Next Dave brings out a collection of mason jars, to show the many ways these three primary-color tobaccos can be blended. First up: Black Frigate, a blend of Cavendish (Virgina that's been steamed during curing), Latakia, and a dash of Turkish. It's a plug cut that resembles a cube of turf or peat, fraying at the edges. Soaked in rum for seven days, Black Frigate smells deep, musty, diesel-edged.[2]

War Horse combines burley, Virginia, and Kentucky—the latter refers to burley that's been "stoved" or cooked in an oven with aromatic hardwoods. It's got anisette top notes that lift the scent upward. Some smokers consider this an aromatic tobacco in that there's a discernible anisette flavor.[3]

Now for Sun Bear. ("All tobacco blends have names like race horses," Dave remarks.) This tobacco looks pretty, unfurling like lacy confetti. It's a blend of red and bright Virginias with Orientals, "cased" with tequila and elderflower. Casing a tobacco differs from flavoring it in that casing is usually unobtrusive, whereas flavoring tobacco gives it distinctive top notes. Casing helps different tobaccos in a blend to meld more seamlessly. During a smoke, you don't want a mouthful of pure Virginia, followed by a mouthful of pure Oriental. In Sun Bear the tequila casing tempers the Virginias' high sugar content, slowing its fast burn, and the elderflower gives that sweetness a floral edge.[4]

He has others: Gas Light, which is mostly Latakia. Byzantium, which is Latakia mixed with Perique, a peppery tobacco used more as a seasoning than a main ingredient. Peterson "Royal Yacht," reputedly smoked in copious amounts by the Belgian writer Georges Simenon as he cranked out the Maigret mystery series. Black House. Billy Bud. Father Dempsey. Jackknife Plug. Old Dark Fired. Night Cap.

You could say pipe smoking is an extended, and very thorough, contemplation of tobacco's smells. It gives you a way to stretch time and luxuriate inside a smell for a long while. This isn't just my notion; the idea of smell appreciation is codified in tobacco blend reviews. Smokers jockey to describe each blend's "tin note": the incipient-rich whiff that's released when you crack open a fresh tin of tobacco. Smokers also talk a lot about a blend's "room note," the scent lingering in a room after a smoke.

Contemplating tobacco's smells continues with the so-called char light, when you first touch a match to a pipe's packed bowl.[5] The char light burns off any excess moisture and prepares the tobacco for a sustained burn. The char light gives you the first whiff of a tobacco's room note.

After the char light you might tamp down the bowl again, light it in earnest, and then smoke. A perfect smoke, according to Dave, is a steady low burn with as few relights as possible. But pipes are finicky and demand a kind of dreamy attentiveness. They won't stay lit on their own. You keep their embers alive via intermittent "sips," stoking the tiny engine puff by puff. Every smoker develops their own idiosyncratic rhythm, adjusting it to the tobacco blend they're smoking, that day's mood, the weather, the company, the cadence of conversation, or their own thoughts.

"Why pipe smoking?" I asked Dave one day. I already knew part of the answer: he'd switched from smoking cigarettes to a pipe initially out of COVID fear. He inhaled deeply in thought; he wasn't smoking a pipe at the time but the gesture itself seemed informed by the habit. "I once heard somewhere: You can't be angry while you're smoking a pipe. That appealed to me in such a deep way," he said. "The peaceable aspect is also appealing. You can't do anything else while you're smoking a pipe. I might read or work on some writing." Pause. Sip. "You forget what you're doing, your sense of time. It gives you a way to fantasize about another world."

Let me now state the obvious: Pipe smoking, or smoking tobacco of any kind, is emphatically terrible for your physical health. But mental health is another matter, sometimes perversely so. During times of unusual stress and uncertainty, the ritual of pipe smoking can be steadying. It fastens one's attention to something small and controllable. For Dave (a painter without a studio who previously smoked cigarettes) it's an aesthetic outlet, a form of armchair exploration he can still do while stuck here.

Smoking a full pipe requires enough free time that you can leave a smoke altered, your mind made up about something. As the scent dissipates and the air clears, the right choice can seem obvious. It recalls how indigenous North Americans—who first cultivated tobacco between 5000–3000 BCE and incorporated its use into nearly every cultural ritual—used smoking to solve problems. They'd choose an issue to resolve, and then ruminate on it for the duration of a cigar or pipe smoke. At the end, they'd toss the ash on the ground. The ash's shape was considered an omen informing their final decision. I like this kind of aleatory decisiveness. Solving knotty problems should depend on rational thought but also on chance. This indigenous practice seemed to respect the limits of how much control anyone can exert over *any* decision they make.[6]

Prior to writing this book I'd never realized how much you can learn by smelling along with friends (or magnanimous experts). Much like other smells—wine (chapter 4), beer (chapter 7), ambergris (chapter 8), truffles (chapter 4)—tobacco smelling is social. Most of what I learned about tobacco's smells was filtered through Dave's self-education. Thanks to him I can explain several causes of "tongue bite," a pipe smoker's scourge. (Beware overly wet tobacco and match your puffing speed to your tobacco at hand. An English tobacco may be smoked more quickly than a Virginia, which is less forgiving.) Thanks to him I can extol a well-executed retro-hale, in which the smoker inhales simultaneously via mouth and nose, briefly fusing tobacco's taste with its smell. Dave persuaded my husband, Seth, to buy his own pipe and collection of tobaccos, and for a time they met every Thursday for an evening porch smoke. He gave Jinn an adorable corncob pipe that she likes to smoke with him. When we all meet for distanced drinks outdoors, Dave is always keen to let us smell new tobaccos he'd acquired, each incredibly deep and loamy and specific.

Each of our friends contains worlds within worlds. Conversation allows us to penetrate those worlds briefly, dwell in each other's shared atmosphere. I will sorely miss our friends when they go, but I'll also be glad they can leave. And when they go, the room note of this time—its humdrum terrors and dreamy boredom, and the deep compensatory solace in friendship—will linger long after.

 **EARTHY**

Dwelling mostly underground, these sections think about terroir,
the mysteries of fighting disease, and smell-secrets lodged deep in dirt.

TRUFFLES

When our airmail package of fresh summer truffles arrived, strong whiffs of the ocean rolled from it in every direction. The scent was surprisingly marine: salty, moody, large in scope, with a kind of insistent depth. Inside our fridge crammed with condiments, leftovers, and canned seltzer, the truffles cast a mesmerizing smell. It was heterodox, suggesting its own strange universe.

I can't say this smell was immediately entrancing. Every time you opened the refrigerator, a veil of intriguing weirdness suffused the kitchen anew. Since we were making homemade pasta that afternoon, the fridge door opened and closed quite a lot. The truffle smell kept shifting with each puff of air that escaped the fridge's interior. As afternoon wore into evening, its bright, fizzy marine note gave way to an earthier, mushroom scent with effervescent top notes, like a glass of sparkling white wine warmed by your hand. Yet that doesn't half describe it. Other mushrooms in my eating experience—which has been pretty extensive—have a paler, more sodden smell, one that points downward toward dirt and leaves. Truffles capture the rich middle note of mushroom smell, then extend that note languorously upward. Truffles take a familiar smell and dilate it.

I opened the bag, pulled out a paper sleeve, then reached inside it. The truffles were wound up in slightly damp paper towels. These were black summer truffles (*Tuber aestivum),* which one expert memorably described online as the Honda Civic of fresh truffles. (Crushing, really: I paid a price for them that seemed Mercedes-level at least. But I was shopping under the gun in late June, and I trusted the deal was fair out of sheer ignorance.) Summer truffles occupy a distant third place to the number-two species, black diamond truffles *(Tuber melanosporum).* These are historically named Périgord truffles for a region in southwestern France, although today most of these truffles hail from farther east near Provence. Far and away the costliest species is the white Alba truffle *(Tuber magnatum)* from the Piedmont and Umbria regions of Italy, named for its ancestral home.[1]

My truffles tumbled out of their paper-towel sheath: bulbous, knotty, like four sizable pebbles or tiny brains hardened in plasticine. Their outer skins were sooty-black and cross-hatched with fine lines, like *craquelin* on French pastries but burned to a crisp. I'd never seen an actual truffle, let alone held one in my hand. They felt hard, like roughened golf balls.

What are they, exactly? Truffles are the fruiting bodies of fungi from the genus *Tuber.* They start as subterranean growths on the roots of certain trees: oak, beech, birch, hazelnut, and pine. Tree and truffle develop a mutually beneficial relationship: the trees, doing their photosynthetic thing aboveground, provide truffles with carbohydrates. Belowground, the truffles help the trees draw nutrients like nitrogen and phosphorus from clay-heavy or chalky soil.

When a tree root is inoculated with truffle spores, this creates an extended network of fine roots around the tree roots, exponentially extending the tree roots' surface area. These networks form natural nitrogen sinks that trap more of this nutrient in the soil surrounding the tree's roots, while also attracting insects and bacteria that further enrich and stabilize the rocky soil. Many truffle-loving trees couldn't thrive without the truffle network's assistance.

Most truffles are harvested in late autumn to early winter, in symbiosis with the tree's natural cycles. In late summer, a truffle may fruit quietly underground in an attempt to propagate itself sexually. (They also reproduce asexually, of less interest to us because it's an odorless, truffle-free process.) As photosynthesis slows to a stop and the tree's leaves fall aboveground, belowground the truffle awakens from its dormant state. Late-autumn thunderstorms and a final blast of heat can both help the truffle along in a fundamentally mysterious process of maturation.[2] As they ripen, truffles emanate a pungent, garlicky, slightly rancid scent that many nut-seeking animals love: voles, chipmunks, squirrels. Female pigs love the scent because it reminds them of androstenone, a mating odor emitted by male boars. While foragers used to train pigs to seek truffles out, they ran into a serious problem: female truffle pigs simply cannot help themselves. In a frenzy of misdirected sexual energy, they dig madly for the truffle, destroying the delicate root networks in the process, and then often eat the truffles themselves. Now truffle hunters prefer to train dogs, who can sniff out a truffle's location aboveground and then simply indicate the right spot to humans, who remove them carefully from the soil.

Black truffles are tricky but not impossible to cultivate. Around 1818—the exact date is far from certain—a French peasant farmer named Joseph Talon planted a stretch of rocky, unpromising terrain with acorns he'd found near truffle-hunting sights he frequented. His goal was simply to cultivate trees to provide some shade. Eight years later, he got his oak trees—which, to his surprise, also had abundant black truffles sprouting from their roots.

Thus began a gold rush of truffle cultivation across southern France. Truffle farming from this period benefited from the plague of phylloxera that killed off many local vineyards, clearing land for truffle cultivation. An epidemic among silkworms destroyed the value of mulberry trees and cleared even more land for an alternative crop. At its peak around 1895, the region produced 1,500 tons of black truffles annually.[3] (The most expensive white truffles have always resisted cultivation. Their relationship to their host trees is cagey and difficult to fathom.[4])

The truffle boom faded in the early twentieth century. Both world wars decimated the European farming population and hastened the growth of cities, leaving truffle fields fallow and uncultivated. Trees only produce truffles for about thirty years, so it took less than a generation for this cohort of trees to slacken production. Since then it has taken decades to rediscover the know-how to cultivate, or even find, these elusive fungi.

we sliced our first truffle with a special device that shaved them paper-thin. Our first truffle was wheat-pale inside and intricately veined. With the truffle's interior revealed, the smell shifted; it reminded me of breadcrumbs crisping on the stovetop, or lightly roasted pine nuts. Together with our friends we prepared four beds of homemade tagliatelle with ample dustings of grated Pecorino cheese and good olive oil. Then we placed the razor-thin truffle over the pasta and set to.

I'll admit it: we weren't automatically beset with reverie. The truffle slices were crunchy on the palate like Catholic communion wafers, not unctuous or as transporting as I'd expected. But is that surprising, really? None of my four diners had had any direct experience eating or preparing fresh truffles. Arrayed around my table were a grown-up girl from Peoria, another girl from Philly (me), a boy from Buffalo, and a boy raised in the Canadian backwoods. We'd all tasted truffle oil before, but this almost never contains actual truffles. The better oils might be marinated with canned truffles, but that's it. Truffle oil usually becomes "truffled" by adding synthetic smells that predominate in actual truffles, chiefly bis(methylthio)methane.[5]

Heat! We decided warming the pasta might unlock the truffles' scent. We heated up more mounds of fresh pasta and tried again. This required using up our second truffle, which revealed a mouse-gray interior and a deeper, more resplendent scent. We shaved this truffle rapidly over the steaming pasta and plunged in again. Now *this* was something. The heat released a fuller range of scents, elusive and changeable, tucked inside the loops of pasta. This time the bower of scents was more capacious and vaulted. Smelling the truffles was a study in microtones, difficult to pinpoint and fleeting, but extraordinary when you did tune in.

Shockingly, we still had more fresh truffles left—and the clock was rapidly ticking on them. Truffles lose moisture quickly, fading in smell, flavor, and monetary value with every passing hour. The next morning, with my rapidly declining assets, I tried some reckless experiments to locate the smell properly. First I microwaved a bowl of olive oil with truffle slices floating in it. Obviously wrong: this killed all the top notes to the smell. Last night's sense of an expansive, complex scent, encompassing everything between dirt and sky, had vanished. Next I warmed a bowl full of good olive oil, not too hot this time, shaved more truffle leaves into it, and loosely covered the pot.

Now I understand why truffle oil has become a Thing. The oil captured so many nuances that had floated wildly around our table last night, so briefly present they felt like hallucinations. True, the oil flattened out the smell's dimensions, too. But it was revelatory in the same way a camera's panoramic view can be. A digital concatenation of photographs won't trick you into thinking you're actually visiting Notre Dame. But that same concatenation will reveal more impossible angles to Notre Dame than you'd ever observe in real life. It's kaleidoscopic, enhanced by direct experience with the real thing.

I shaved the rest of my fresh truffle into a jar of olive oil and observed the smell deepen over the days it took to consume it. The smell never regained that rounded, arching quality, but it hovered over the oil, a beautiful ghost of itself.

WINE

Other than "grapes," here are the first smell-notes I detected in my glass of Brunello di Montalcino: a whiff of self-consciousness, mingled with imposter-scented fear. I felt unexpectedly nervous. In writing this book, hadn't I been pushing myself to observe many different smells and see what stories I could find inside of them? I worried about wine as a proving ground of my nascent skills or lack thereof.

Turns out all this handwringing was needless. Smelling and learning about wine validated the original hunches that motivated me to write this book: that stories, even whole histories, can unspool from different smells. That it's worth getting acquainted with objects via smell more often than we do. That you can improve your smelling skills simply by applying greater attention to them. That smells can matter deeply to people and that talking about them together can make us feel like we're probing each other's thoughts. Wine enthusiasts have gotten, and thoroughly absorbed, all these memos. It's too bad that a fear of snobbery prevents more people from trying to smell their wines. In fact, you can observe a lot via smell even in moderately priced wines, and it makes every glass you drink that much more interesting.

"Smelling wine" seems a bit incorrectly stated. Obviously smelling and tasting wine go hand in hand, and tasting wine would seem to trump smelling in terms of importance to the experience. Or does it? The first surprise of observing wine more closely is accepting the strange, inverted relationship of taste to smell.

Smells are what impart most of the flavor to wine—which is why wine experts think about smells both constantly and systematically. Some of their better-known mannerisms, like swirling wine in the glass, are aimed at opening up a wine's aromatics and releasing its smells more fully. Another way to do this: hold the glass level to the floor, stick your nose way inside and waggle it around a bit. Sniff around that surface tension of liquid like a bloodhound seeking a body underwater. Open your mouth while you sniff, too, as if you're breathing in the wine; this encourages smell molecules to travel retro-nasally over your tongue.[1]

What should you be smelling for? First, check whether the wine is still okay. Sniffing the cork may seem like an affectation, but it's an easy way to confirm the wine hasn't oxidized. Sniffing the cork can also alert you to signs of "cork taint"; a chemical called TCA can infiltrate a bottle via the cork and give wines off-flavors of damp cardboard or gym socks. The cork shouldn't be crumbly or bone-dry, and it should smell like cork and nothing else.[2]

Assuming the wine isn't spoiled, what else should you smell for? Anything that focuses your senses and draws you deeper into the glass. Smelling wines carefully can also reveal clues about how the wine was made.

The process of winemaking is fairly simple and constant across all wines. It's the specific decisions vintners make at each step that change the wine's character and manifest later in its scents and flavors. A proxy for what makes a "good" wine is simply whether it demonstrates appealing variations across different years and bottles. Mass-market or inexpensive wines are produced according to more industrialized processes designed to minimize variation and yield consistent results. For these wines, every year and every bottle should taste the same on purpose. This isn't necessarily "bad" but can trend toward the lowest common denominator of flavor profiles.

Back to process. It starts when a winemaker chooses to grow a particular grape in a certain spot. (In this example, we'll follow the process for a wine made with a single grape type, or varietal. But don't forget that many wines are blends of multiple varietals.) That geography's climate, the slope of land, that year's weather patterns—from the amount of sunshine to temperature shifts—all these factors influence the grapes as they grow. Vintners may also deliberately stress their vines to influence their qualities: depriving the vines of water, for example, or pruning away some clusters to concentrate flavor in the grapes that remain.

The vintner decides when to harvest the grapes, a decision informed by numerous factors as well as qualities they're seeking to produce in the glass. They remove all the grapes' stems and, for white wines, also the skins. (Contrary to popular belief, white wine doesn't come from green grapes and red from purple grapes; it's the removal of grape skins that makes a white wine.) They mash the grapes into a slurry called must and transfer this in a fermentation vat. The vintner then either adds yeast or allows the grapes' natural yeasts to get to work, converting the grapes' sugars into alcohol and carbon dioxide. Some wines undergo just one fermentation round, others do it twice.

The winemaker chooses a strategic moment to stop fermentation in the vat, then transfers the wine to age into casks (either wooden or stainless steel).[3] Wooden casks impart tannins to wine, that astringency that dries out your tongue. Putting a wet tea bag on your tongue will give you a hit of pure tannins, so you'll recognize that taste when it's present in wine. Casks can also impart flavors of the wood they're made from, or from other aromatic liquids previously aged in those barrels—say, sherry, whiskey, or simply older generations of wines.

After cask-aging comes bottling. A few other variants might happen at this final stage. For instance, Champagne and effervescent wines get a teensy bit of yeast inserted under their corks so they undergo a second fermentation inside the bottle, producing their signature bubbles. Some wines age longer in the bottle before getting released to drinkers worldwide. And that's the entire process.

You can learn all kinds of things about a wine from its smell. Experts like sommelier Alex Ring start by memorizing which smells go with which grapes—Syrahs smell like black pepper, for instance. In identifying the grape (or grapes), you might think about what kinds of fruit smells you notice and their qualities. It's in those qualities that other clues—about geography, aging, or process—spool up to the nose. A whiff of dried fruit might tell you a wine is older, or that it grew in a super-hot climate where the grapes shriveled on the vine. Tarter fruits might indicate a cooler climate where the grapes ripened slowly; the vintner probably raced to harvest the wine before the rains came. Woody smells hint at how the wine may've been casked. A burning sensation in your nose suggests higher alcohol content. A whiff of toast might tell you about lingering yeast in the wine. Alex found smells of green bell pepper revelatory as he sharpened his own skills. These indicate pyrazines, volatile compounds prevalent in Cabernet Sauvignon and Sauvignon Blanc grapes. Some grapes have to push past the pyrazine stage as they ripen, so detecting these lingering scents can tell an experienced nose a lot about how the farming went down in the later stage. Some beginning tasters bring samples of typical smells—actual berries, green apples, lemons, pineapples, pears, lychees—so that you can sniff the wine, then sniff the real thing and compare. Your nose, and you, will pick out smells more readily over time.

My Brunello di Montalcino, a 2014 Carpazo available for $25, is made only with Sangiovese grapes. This is the same grape used to make Chianti, perhaps *the* most mass-market red wine going. Sniffing it, you get notes of dried sour cherries—an observation

EXERCISE

Get closer to the ground.

Don your best dirty-wear clothes and then head to the woods, ideally in spring after a rainfall. Find a clearing, then lie on the ground on your stomach. Wearing gardening gloves to protect your hands, scuttle about, sniffing as you go. If you attract unwanted attention, tell passersby you are replicating a UC Berkeley smell study in which participants—blindfolded, earplugged, gloved, and knee-padded—were asked to track a drizzle of chocolate essential oil using only their sense of smell. (Two thirds of the participants were successful in their first three tries; the others showed measurable improvement over time.)*

Don't be shy about overturning rocks, sticks, or sodden leaves. Interesting smells will probably lurk there.

Make sure to compare a particular spot's smells from different altitudes. Sniff with your cheek to the ground, then crouch, then clamber up on your knees, then stand upright. Smell molecules obey gravity: they're buoyed by turbulent airflows but eventually settle near the ground. You can observe this fact yourself by bringing your own essential oil, drizzling it on the ground, and then sniffing from various heights.

---

* Erika Check, "People track scents in same way as dogs," *Nature*, December 17, 2006, doi:10.1038/news061211-18, https://www.nature.com/news/2006/061211/full/061211-18.html/.

I confirmed by chewing a few actual cherries. A tang of aged balsamic vinegar. It's slightly herbaceous, less fruity, and more vegetal.[4]

Our senses always work together, and wine sniffers take full advantage of this fact. A clear, bright-red wine might wash lucidly in a glass and smell of fresh strawberries: these sensory clues taken together suggest Pinot Noir. An inky-red wine might ooze with splendid slowness in the glass, redolent of cooked prunes: all of these clues point to Malbec, Merlot, or Cabernet Sauvignon. Tannins drying out the tongue also give the wine a certain texture; sometimes they smell of wood. My Brunello edges toward darker red in color and rinses clear as it moves around in the glass; it almost looks like balsamic vinegar. Its taste is remarkably spicy, and its tannins pucker the tongue.

Blind tastings are how sommeliers sharpen their skills, and it's an extremely social process. A bunch of people gather to taste wines without knowing which wine is which. They all sip together and take turns verbalizing what they observe. The goal is to glean as much about the wine as you can: grapes, region, appellation (a protected name under which a wine may be sold, like Champagne or Brunello di Montalcino), year, and producer. Blind tastings are rigorously grouped by expertise level; like tennis round-robins, everyone prefers to taste with people better than themselves. Those who advance in the blind-tasting world get more systematic, both in how they narrate their observations and in which tastings they'll attend. An advanced blind tasting might focus only on Grüner Veltiners from a specific region in Austria, or mastering an appellation like Brunello di Montalcino: wines made from specific grapes from a certain district according to a well-defined protocol of rules.

During the pandemic while writing this book, I couldn't participate in any live tastings—but I enjoyed the next-best thing, a community of wine fans smelling, tasting, and observing smells all across the internet. I Zoom taste-tested with my father-in-law, Howard Spiegelman—a wine instructor and Brunello fan. When a tasting is really clicking, observing wine smells with other people can feel like an improv troupe finding their flow. It's all *Yes, and*. You can actually smell and taste different things once a more expert palate has pointed them out. It's intimate and sweet but demands a blend of insouciance combined with focus: the fun of taking something more seriously.

BACK TO ME AND THE GLASS AND THE SMELLS. Previously the feeling I brought to smelling wine was one of incipience—a quivering arrow of anticipation, pointing straight at the future. Time to drink something yummy! But now I can see how this smell's vectors proliferate, pointing in many directions at once. The smells I observe right now arc back to the past, revealing how the drink was made. They connect me with other people, all of us sniffing together, a momentary closeness between friends. And perhaps the most important vector isn't an arrow pointing anywhere. It's the dot of *now*. A wine smells the way it does because of innumerable factors: how it was made, how it aged and changed over time, what I bring to it, what my friends notice in it. It's even changing in the glass between sips, breathing and opening up. It's almost as alive as I am.

CANNON FIRE

What *does* cannon fire smell like, anyhow? Other than military historians, only hunters, fireworks technicians, and Civil War reenactors can accurately describe it. Gunpowder recipes have changed over time, but its bouquet has remained consistent: rotten-eggy, sulfuric black powder as the basenote. The urine-like tang of saltpeter. The particulate density of charcoal.[1]

It must smell kinetic. Cannon fire fills empty air with mass, boom, taste, and grit. It roils and churns the atmosphere. If a street scene pixelated into rubble before your eyes, it'd look cinematic, unreal. But smells and dust clogging your nostrils would be palpable and undeniable.

IN THE AMERICAN CIVIL WAR, advances in mechanized warfare automated killing and maiming to produce a war bloodier than either side had ever imagined. In his book *The Smell of Battle, The Taste of Siege* historian Mark M. Smith conjures the human toll of this war's carnage via smells as well as the other senses. In their letters home, Civil War soldiers struggled to convey the sensory bombardments of the modern battlefield— and smells ranked highly in their reminiscences. "Smelling the powder" became military shorthand for fighting on the front lines, for witnessing action up close.

Cannon fire was both the first and last smell of war. It initiated every scene of battle and then lingered long after the fighting ended, connecting the impossible dots between "before" and "after." Picture a farmland plot drowsy with rural smells—cow manure, fresh-mown hay—suddenly erased by the smoky, nostril-stinging cloud of cannons. As the fighting commenced in earnest, cannon shot both filled the nostrils *and* the ears. Cannon fire "whirred over our heads like a shower of bumble bees," wrote one soldier. Heavy cannon balls would trace a slow auditory arc one could sidestep by keeping one's ears perked; bullets and shells were silent and therefore less easily avoided.

"The pungent smell of burnt gunpowder pervaded the evening air," relates one contemporary account of a battle's aftermath. "The air, thick with smoke and sulphurous vapor, almost suffocated the troops." Acrid smoke mixed with stenches of carnage compounded by the sheer number of dead awaiting burial. After visiting Gettysburg, the war's bloodiest battle site fought during a July heatwave, one woman's diary entry reads: "The atmosphere is loaded with the horrid smell of decaying horses and . . . it is said, the bodies of men imperfectly buried. I fear we shall be visited with pestilence, for every breath we draw is made ugly by the stench."

After fighting ceased cannon fire's smell kept triggering memories, both patriotic and PTSD-like, in surviving soldiers. One Philadelphia reporter described Fourth of July fireworks a year after the Civil War ended: "Not until the weesma hours's of the 5th

The urotoxic coefficient is at least a fourth higher in Germans as in French. Thus, if 45 cubic centimeters of French urine per kilo are sufficient to kill a guinea pig, the same result can be achieved with only 30 cubic centimeters of German urine . . . [thus the German's] overworked renal apparatus is unable to evacuate all the uric matter from his body and . . . some of it must be voided through the plantar region. There is thus a physical and factual basis for the saying that the German pisses through the soles of his feet.*

—Dr. Bérillon, World War I physician

* "La Bromidrose fétide des Allemands," in *Bulletin et mémoires de la Société de médicine de Paris,* 1916, 142–45.

will the atmosphere cease to be ladened with the fumes of the villainous saltpetre!"[2] Other ex-soldiers sniffed the gunpowder vigorously and chuffed themselves for a job gloriously done on the battlefield. Cannon fire was a smell mutable with your politics, with how firmly you believe war can solve anything.

CANNON FIRE HAS BEEN DEPLOYED to fight enemies both visible and invisible—for instance, airborne illnesses. Let us now introduce miasma theory, the belief—prevalent until 1880—that bad smells literally caused disease. Miasmas were believed to leak from garbage heaps, privies, hospitals, and poorhouses—but also from cemeteries, swamps, caves, paths heavily shaded by trees, even cracks in dirt sidewalks.[3]

Miasma-fighting theories are amusingly wrong-headed in a way only debunked science can be. Annick Le Guérer's book *Scent* traces miasma theory's false logic. Yes, most infectious diseases spread via invisible means. Yes, where sweet-smelling cleanliness prevailed, illness also tended to abate. Yes, as cities got overcrowded, illness and stink concentrated in denser areas, too. But fighting rotten smells was, at best, a rotten approximation for fighting actual disease.

According to miasma theory, one could protect oneself from less lethal miasmas with good ventilation and careful personal habits. "In order that the primal qualities of the air and body should remain stable, care had to be taken that neither would become too damp, too hot, too dry or too cold," writes Le Guérer. During a 1348 plague in Paris, the Faculté de Paris advised residents to breathe cooling perfumes of rose water, camphor, and sandalwood in summer. In winter, one should sniff aloewood, sweet gum, and nutmeg scents. Bathing too frequently in any season, they reasoned, might endanger one's health. Better to keep your skin's pores plugged protectively against infiltrating disease.[4]

Some scientific factions favored fighting miasmas with pleasant smells like scent boxes, cigars, lozenges, syrups, or clutching a tiny, fragrant citrus tree.[5] One sixteenth-century text advises physicians to approach patients armed with a juniper branch and a pomander ball. (One reason stethoscopes gained physicians' favor was because they enabled patient examinations from a healthy distance.) Other practices advised during epidemics included compulsively sniffing red carnations, carrying lemons studded with cloves, and sprinkling one's clothes with pulverized angelica, a common herb in northern Europe.[6]

Some doctors argued for fighting miasmic smells with even worse smells. In 1622 Dr. Jean de Lampérière encouraged physicians to protect themselves against disease by rubbing their naked bodies with a sooty black mixture containing dried peacock dung and goat urine.[7] An outlandish recommendation, and yet not totally amiss: the smell of goats (and cattle, sheep, and camels) does repel the fleas and ticks that transmit bubonic plague.[8]

Plague fumigations by cannon reached their zenith in seventeenth-century France. Professional perfumers would decontaminate whole households after plague victims were carted off. Wearing pans of lit charcoal hung around their necks, perfumers ignited a bonfire in front of the house and shut all exterior windows. Then they'd set to work indoors: gutting straw mattresses, carrying off dirty linens to hot ovens, wiping down furniture with wine and vinegar, dipping jewelry and silverware in boiling water, maintaining aromatic bonfires in the granary and storerooms, and on and on. The checklist was vigorous, extensive, and always pungent. Meticulous perfumers would crown their work by firing cannons in the streets, to "dispel the infection that may linger in the woodwork or on the outer walls of buildings," according to one contemporary report. Cannon firing caused problems of its own, like rupturing building foundations and breaking windows.

The fight against nasty odors had durable effects. Fighting miasmas—and noxious urban smells, in the modern era of germ theory—shaped urban landscapes in numerous ways. Decisive moments of reek, like the Great Stinks of London (1858) and Paris (1880), transformed metropolitan areas.[9] To keep city air cleaner and fight disease transmission, lawmakers paved sidewalks and whitewashed walls, engineered sewer systems, established health boards and zoning ordinances, introduced environmental reforms, widened the streets, and planted public gardens as "urban lungs."

So the smell of cannon fire contains multitudes. In centuries past when it dominated wars, it reeked of carnage. Threading through premodern city streets, this smell registered differently: like the avenging angel of health and sanitation. Today cannon fire smell appears rarely, only in spent fireworks. But that's a fitting transformation, really: what used to smell of warfare and a futile form of pestilence fighting now smells like civic pride and the triumphs of modernity.

MELTING PERMAFROST

This is one of the few smells in this book I won't experience directly. So getting inside requires closing the eyes and focusing, an exercise of imagination.

I watched a video of Arctic researchers describing their impressions of this smell. Grassy. Peaty scotch. Musty. Stinky feet. One researcher likened it to walking into a pine forest and stumbling on a pile of manure. Another compared it to unpacking a muddy tent that's been stowed away for the season. Aged, good-quality cheese. A compost pile. Many remarked how there's something very old about it. One researcher noticed hints of maple syrup and flower undertones. Mold in a damp room. Smoke. Like the crawlspace in a dirt basement, with a distinctly metallic ping.[1]

That metallic ping is what extraordinarily high concentrations of CO_2 smell like. And that's the terrifying idea unlocked with this smell: permafrost should be odorless, because this soil should've always remained frozen.

Permafrost covers a quarter of the land in the Northern Hemisphere, and ordinarily it never thaws. The soil above it warms enough for tundra plants to grow and Arctic animals to roam over it. During their brief lives, those tundra plants breathe in CO_2 just like other plants and thus extract carbon from the atmosphere. But when Arctic plants and animals die, they don't decompose as they would in more moderate climates. Even if they die at the peak of midsummer, winter is only a few short weeks off. They lie exposed in lightly refrigerated air until they're buried under fresh waves of snow. Gradually they settle deeper and become carbon-rich deposits in the permafrost layers beneath. This process has run naturally for millennia, locking up lots of carbon in permafrost.

Not just "lots." The amount of carbon locked inside permafrost is *double* the amount of carbon already free-floating in our atmosphere—an amount we already know is unhealthy for the planet and increasing too rapidly without the compounding acceleration that melting permafrost represents. And melting permafrost releases not just the smelly burps of carbon dioxide and methane into the air. It also releases stealthy, odorless threats: The largest repository of poisonous mercury on earth. Microbes of largely eradicated diseases like smallpox, anthrax, Spanish flu, bubonic plague.

Everyone knows we need to have mud for lotuses to grow. The mud doesn't smell so good, but the lotus flower smells very good. If you don't have mud, the lotus won't manifest. You can't grow lotus flowers on marble. Without mud, there can be no lotus.

—Thich Nhat Hanh, Buddhist monk and peace activist

In other words, this smell is dark. Like the smell of burned gunpowder, it's a smell of something that should never have happened. And yet it's not all unmitigated terror—some shots of light penetrate the darkness. Permafrost also smells like the deep past, many ancient layers of it, coming to light. It smells like extinct animals, early human encampments, illuminating discoveries. Because no living thing fully decomposes in the Arctic, an era's worth of flora and fauna can reemerge whole. But when they hit the warming air of modernity, decomposition begins anew—unlocking antediluvian microbes from their cryogenic freeze. Some will reanimate, with unpredictable results.[2]

Permafrost also smells like other volatile organic compounds (VOCs) besides carbon and methane. Highly diverse and as yet not widely studied, these VOCs permeate the tundra above permafrost with both bad and good effects. They're consumed greedily by microbes living in the tundra and, if they penetrate even farther upward, can react rapidly in the atmosphere, producing ground-level ozone that's toxic to both forests and humans. But those same reactions offer a silver lining: they also spawn tiny particles that scatter solar radiation back into space, where it forms into clouds and cools the climate. It unleashes a domino effect with proliferating consequences. Some whiff of that imagined smell is hopeful, even if it's a wan hope.[3]

I was sorry not to smell this for myself. But knowing what I know about it now, I wonder: could I have even borne it?

TEA

It's inviting, this smell, suggesting great depth. A whiff of woodsmoke lifts from my tea leaves, a smoky black Lapsang Souchong, with a hint of pine resin. It smells syrupy and rich like raisins, welcoming and combustible like tobacco. Smelling the dried leaves draws you briefly into melancholy. It's an outdoorsy, all-by-your-lonesome smell. The scent contains within itself a picture of waiting: you, alone in the cold, awaiting a hot drink to buck you up. Sniffing the leaves again, one thinks suddenly of citrus zest—a glint of sunshine slicing through the moodiness.

Brewed in a cup, the Lapsang Souchong changes. Its discrete notes meld into smoothness. It tastes full, lucid, entirely without bitterness. Its nuances are still present, just resolved into balance. All the skill it took to harvest the tea leaves blooms in the cup just by adding boiling water.

Tea is dead simple and therefore complex. It delivers caffeine, a major selling point, and it can warm us up or soothe us down. But tea drinking offers no nutritional value. Drinking tea isn't necessary; it's beautiful and mostly useless. We drink tea solely for the tastes, smells, and ritual of it. Because it requires piping-hot water, tea forces you to wait. Tea slows down and dilates the moment of smelling something lovely. And it gives that smell a physical presence, filling a mug.

We call this drink "tea" now but it started out as *t'e*, a Chinese Amoy word pronounced *tay*. The Dutch first imported tea to the West from the Amoy port in Fujian Province; they called the drink *thee*, giving rise to the English word *tea*. The Mandarin Chinese word for tea is *cha*, which morphed as it traveled the globe—through India, Persia, Afghanistan, and Russia—into *chai* and *shai* in Arabic.[1]

Tea began as a practicality. In the third century BCE Emperor Shen Nong decreed that citizens must boil their drinking water to purify it. Boiling water himself one day, as the legend goes, some tea leaves blew into the cup and brewed with it. Shen found the resulting drink invigorating and delightful. These were the original tea leaves from the plant we now call *Camellia sinensis*.[2]

There are six kinds of classic tea. White, yellow, and green teas are all unfermented. Fresh leaves are picked, then dried or steamed immediately to prevent oxidation. Oolong, the fourth kind of tea, is semi-fermented; black and pu-erh teas are both fully fermented. Pu-erh tea is rare outside of the Yunnan Province of China, where tea leaves grow on thousand-year-old trees on remote outcroppings of six mountains. Early legends tell of agile monkeys gathering pu-erh tea leaves from these trees. Among all tea types, pu-erh tea distinguishes itself with its earthy aromas and flavors.[3]

Early Chinese teas smelled bitter and medicinal. A dictionary from the Wei dynasty in the fourth century CE advises placing dried tea leaves loose in a pot, pouring boiling water over them, and then adding onion, ginger, oranges, and possibly salt.[4] Only later did tea makers start scenting tea leaves with aromatic florals and herbs, like jasmine. The process of scenting tea is simple: tea leaves are piled up next to fresh flowers to absorb the latter's fragrance.

The world's first treatise about tea is called the *Ch'a Ching*, written by poet and scholar Lu Yu in 780 CE. Yu addresses every facet of his subject with wit and thoroughness. He describes the moment when boiling water hits the tea leaves, what one sees as the tea's perfume starts rising:

> Tea has a myriad of shapes. If I may speak vulgarly and rashly, tea may shrink and crinkle like a Mongol's boots. Or it may look like the dewlap of a wild ox, some sharp, some curling as the eaves of a house. It can look like a mushroom in whirling flight just as clouds do when they float out from behind a mountain's peak. Its leaves can swell and leap as if they were being lightly tossed on wind-disturbed water . . .[5]

Because brewing tea requires very little equipment, the practice lends itself to becoming conceptual and aestheticized. Tea making is a mirror for human complexity, our collective and individual urges to make something perfectly, to consider all its angles. Tea distills culture.

The Japanese started drinking tea in the late sixth century CE thanks to Buddhist monks and priests traveling to China to study, then returning home with supplies of tea. The formal Japanese tea ritual, known as *chado,* "the way of tea," arose in its current form in the sixteenth century.[6] (To learn more about *kōdō,* or the "way of incense," see the oud section in chapter 5.) Traveling across the Asian continent, Tibetans make an extremely strong brew they call "butter tea"; it calls for yak milk and salt and treats the tea like a vegetable—which for them, living at high-altitude where vegetables don't grow, it is.[7] A special Afghan tea, called *qymaq chai,* is made from green tea flavored with cardamom. Adding bicarbonate of soda turns the tea red; served with copious amounts of sugar and milk, the final beverage is pink.[8] Now reserved for formal occasions, it's a vestige of the drink Silk and Tea Road traders served to seal their deals. The Russians contributed to tea history the samovar, a device that spread to Turkey and across central Asia. Samovars keep brewed tea hot and plentiful; their presence encourages drinking multiple cups and fills awkward pauses in conversation with the chance to refill. Russians serve tea with a slice of lemon and a spoonful of fruit preserves.[9] Moroccans brew green tea with fresh spearmint and sugar; they pour it from a silver pot from a great height to froth the drink,[10] a ritual that spread across North Africa.

Tea reached Europe—specifically the Netherlands in 1610—via Dutch shipping routes from the East. Tea drinking wasn't an intuitive practice for some Europeans. In an anecdote from 1685, the Duke of Monmouth's widow sent tea to her cousin without explaining how to prepare it. The cook boiled the tea, threw out the water, and then served the wet tea leaves on a platter like cooked spinach.[11] Yet tea handily conquered England by the mid-eighteenth century. At first tea was expensive, but prices dropped with the Commutation Act of 1784; by the mid-nineteenth century cheaper Indian tea was being imported from plantations across the British empire. Brewing it purified iffy drinking water, and unlike coffee, tea was easy to prepare. Afternoon tea became a daily British fixture by the 1850s.[12]

Tea crossed the Atlantic with American colonists—who famously resisted British taxation, dumping tea into Boston's port waters in protest. Americans savored tea much as the British did, contributing the innovations of iced tea and disposable teabags in the early twentieth century. Elsewhere in North America, the Innu people of Labrador, Canada, used to migrate to seasonal hunting grounds and carry all their belongings with them. Their children carried "tea dolls" stuffed with tea leaves. They'd use the tea, then replace the doll's stuffing with grass or hay.

Tea in the Australian outback was prepared in a rambunctious way. You suspend a "billy can," or metal bucket, over a fire, heat the water, brew the tea good and strong, toss in a gum leaf for extra flavor, then swing the bucket vigorously around your head to settle the tea leaves to the bottom. The final drink was served with heaping portions of sweetened condensed milk.[13]

Everywhere around the globe, the goal of making tea is the same: to perfume water and then to commune with it. It resembles distillation in perfume to an almost literal degree. Tea's scent is inflected with many different top notes—mint, milk, salt, cardamom, lemon—but the base note is fairly constant. That communality is itself comforting. Tea gives you a bracing pause, to sip and breathe it in. It turns a smell into a fortifying moment.

 # RESINOUS

These smells all derive from wood, plumbing concepts as lofty as luxury and worship and as humble as childhood memories.

FRESHLY SHARPENED PENCILS

OUD

CAMPHOR

FRANKINCENSE

MYRRH

FRESHLY SHARPENED PENCILS

A warm, all-encompassing wood. Each inhale dries out the nostrils' insides more. The smell is dense, particulate, and airless, like the inside of a cupboard. The tail end of the smell pings metallic, but distantly so.

Somehow, the smell seems shrunken compared to childhood memory. Why?

If writing has a smell, it's a pencil's smell. This is my stubborn and unshakeable conviction, despite the fact that I'm typing this chapter on a computer. Pens are lovely but too permanent to represent writing as a live process. (Come at me, haters!) Pencils accept the realities of mistakes, erasure, palimpsest. Writing, if the activity could take material shape, would always manifest as a pencil.

Yet a pencil's status as a physical object is slight. Pencils eat themselves, as do erasers. You write and write, laying down a weightless ribbon of graphite. Trailing in that ribbon's wake are heaps of pencil shavings. Most of a pencil's matter—and, as it turns out, its scent—gets thrown away.

I fill a glass jar halfway with shavings, then stick my nose deep inside it. Still I don't smell much. I toss the shavings like hot cider heavy with sediment, then sniff again. Eventually I tip the jar forward as if my nose is thirsty. (If you want to smell something deeply, bashfulness does not pay.) The shavings crowd around my nose, tickling in a friendly way. From inside this thicket, the smell spools outward.

How closely does *this* smell match the pencil smell from my childhood? Have pencils always smelled this way?

"Memories, even if they go broadly, don't always represent an autobiography," wrote Walter Benjamin in his 1932 essay *Berlin Chronicle*.[1] "Autobiography deals with time, with processes and sequences, and what constitutes the continuous flow of life. Here I'm talking of a space, of moments and discontinuities." Benjamin saw memory as spatial, labyrinthine, inwardly braided, not linear.

Like remembering, smelling also feels like entering a tiny space. A smell opens rapidly like an enfolded structure, presenting all its angles at once. It's a dense space of unreliable distances where one can roam only briefly.

Each time you sniff something, the moment bristles with data—all poised to vanish when the nose overloads. It's disorienting. Fortunately, the act of smelling is repeatable. You just take breaks and start again. Let the data fluctuate, register, fade. Observe whatever you can. The space of smell, like memories, will always reopen.

AN INANIMATE OBJECT'S VERSION OF MEMORY IS ITS HISTORY, the story of how it came to be what it is. The history of pencils is slow-moving in the extreme: they've changed only incrementally over the last few centuries.

Smell offers a kind of deep information about an object. After all, to smell something is to ingest its particles via your nose. Yet we don't usually acquaint ourselves with objects this way. I wanted to smell every inch of a pencil—eraser, wood, and graphite—and see what I could learn about pencils by this closer examination (augmented with a little research).

Here's the first surprise: pencil wood used to be much more fragrant. The first modern pencils were made of Eastern red cedar from Florida, Georgia, and Tennessee. A smell that still wafts from cedar closets, red cedar made pencils that smelled of spicy black pepper and cinnamon and colored shavings pinkish-red. As late as 1890, red cedar was so abundant Southerners built barns, log cabins, and fences with it—that is, until the production of millions of pencils thinned the wood supply and jacked up prices.

Thus began the search for new pencil woods. After considering numerous species, the U.S. Forest Service finally recommended incense cedar from Oregon and California. This wood was cheap and functionally ideal for pencils. But back then pencils made with incense cedar—the pencils we have now—were considered too pale and too weakly scented to please consumers.[2]

So early twentieth-century manufacturers dyed incense cedar pencils and perfumed them to simulate red cedar. History doesn't record when manufacturers stopped this practice. Perhaps they thinned the perfumes and dyes to cut costs, and customers didn't squawk. So they thinned them more and more, and so the smell of pencils dwindled to almost nothing. It's bracing to consider: my brand-new pencil shavings might smell musty and stale, but century-old shavings would still be pungent, spicy, and new. It flattens time somehow.

What about the graphite—could I *really* smell in it my jar of pencil shavings? I take the little stick inserts packaged with mechanical pencils and sniff those.

The smell is clean, bright, forward, uncomplicatedly metallic. It's almost a shrill smell by itself, but not unpleasant if you imagine it mingled with metalwork-shops smells: warm lubricants; keener solvents; metals at rising or falling temperatures; sweet grease.

Pencils weren't always graphite encased in wood. The earliest pencils were sticks of lead alloy wrapped in string or paper. Graphite replaced lead alloy in pencils in the sixteenth century, with the discovery of an abundant graphite supply in Borrowdale, England. Graphite produced a darker mark than lead, yet still erased cleanly.

How does graphite get inside a pencil, anyhow? The original manufacturing process required grooving a long strip of wood, then inserting whole graphite slats into the groove. Cover this with a square wooden top, glue the two pieces into a graphite-wood sandwich, then sand off the square edges to produce a cylindrical pencil.[3]

Patented in 1795, the Conté method allowed manufacturers to make quality pencils with lower-grade graphite powder. First you remove the graphite's impurities, mix the powder with clay and water, dry it, and then fire it in ceramic. This fuses the mixture into a graphite slat. Insert slat into wood casing, glue closed, sand, paint, repeat.[4] Adding clay and wax not only stretched the supply of useable graphite, it also gave pencils an additional marketing feature consumers like: variety. Adding more clay and wax yields a sharper, pale line, whereas a pencil containing more graphite writes darker and softer.[5]

Now for the punch line: it turns out pure graphite has no smell. Whatever I'm sniffing is just clay and wax additives. But you could say this whiff conveys something more historically important that mere material: I'm smelling an early triumph of the Industrial Revolution.

Erasers reveal themselves in smell, too. Quality erasers contain more natural rubber, both the best material for erasing and a highly aromatic one.[6] An odorless eraser is usually a cheap and useless one. Natural-rubber erasers smell cheerful, ugly, forthright.

I sniff my erasers unnaturally—curling a rubber oblong under my nose, like a mustache whose tips grew aggressively into my nostrils. But normally you'd smell an eraser only while using it. The act of erasing makes us hunch closer to the paper. To blow away the twisted rubber strings, you must first inhale, and smell exactly what you're doing.

A pencil's smells emerge not while writing, only in sharpening and erasing, in the pauses between. You could say pencils smell like incense cedar, clay, wax, and rubber. But you could also say: pencils smell like thinking.

PENCILS ALSO INTEREST ME because they're a rarity: a smell whose first impression I might be able to recall. People don't wave pencils under a baby's nose. I probably first smelled pencils when I started drawing—when I started remembering anything at all, in fact.

Smells predate the formation of actual memories by a lot. When you encounter a brand-new smell, it hardwires associations into the brain. Overwriting that first impression of a smell later is difficult to impossible.

As we saw in part 1, The Nose, smells enter your nose and ultimately reach the olfactory bulbs, where they're processed. Unlike other sensory stimuli, smells bypass the brain's thalamus, a central relay station that handles complex, higher-order processing and most of our other sensory perceptions. The thalamus is part of what's known as the "new brain": evolutionarily more recent, these structures govern brain processes that tend toward the reasonable, the civilized, the coolly executive. As smells sidestep the thalamus, they zap straight from the olfactory bulbs into two old-brain structures, the amygdala (which regulates emotions) and the hippocampus (which controls episodic memory). Smell is decidedly old brain: aloof from logic and language, emotionally raw, a palimpsest for life memories. Smell transports us instantly into a past scene, rolling the cinematic tape complete with feelings and a sense of embodiment that can be upsettingly "live." No wonder we sniff with some hesitancy.

Could I recall my earliest contact with pencil smell? What other details might emerge, stronger and clearer, with that bygone sensation? Back in the twenty-first century, I sniff my pencil shavings again and remember my way back into childhood.

I'M STUCK AT A DESK, one among thirty bent heads, getting it done: worksheet after worksheet. Bubble, bubble, bubble. Sharpening a blunted pencil gives me an excuse to stretch my legs, wander the classroom.

I head to the back corner where the pencil sharpener hangs and other smells collect: dank rubber boots in the back closet. A rust-laced whistle as steam radiators awaken. Chalk dust sweetly reminiscent of confectioner's sugar. The pervasive dryness of rooms filled with paper.

Sharpening a pencil is a mostly invisible process that activates all the other senses. Bracing the sharpener's sides, I feel its grinding gears and yielding wood. I hear the pleasant chewing sound.

The sharpener's cavity fills until it's time for emptying. I scoop out the yellow-edged wooden curls, brush away the glittering soot from its crevices. That's when the smell punches forward.

I sniff my shavings jar again and remember my smaller body occupying that corner. I remember boredom. Scribbled notes passed to buddies. Wild daydreaming. The blank, wide canvas of the unimaginable future. Sniffing my pencil now, the smell amplifies almost magically at these thoughts. It's like tuning into a crisp radio frequency in the middle of nowhere. It conjures a vanished civilization.

OUD

The word for this smell is *plush*. Deeply hospitable. Imagine entering a cedar closet—dry and spicy, mildly suffocating yet bracing—and finding it transformed into a Fabergé egg. Or standing in a forest in which all the trees happened to exhale perfumes. This is nature transformed by the polishing alchemy of culture.

Oud is a deep, fabulously expensive aromatic wood. Its original base note is wood in the same way that polished mahogany or delicate inlaid carvings are, technically, wooden. The smell goes beyond "warm" as a descriptor; it almost seems to generate its own heat. Oud is a baroque smell with an ample, long-lasting *sillage*—a term referring both to a perfume's staying power and the scent trail lingering in the wearer's wake. It cloaks the wearer protectively, like a nimbus or a mantle.

In a word, this smell is maximalist. Rich and fully dimensioned, oud's pleasures come from surrendering to a scent dizzyingly rich in detail. Notes emerge differently with each sniff: woody, balsamic, grassy, powdery, brightly sweet. If this smell had colors, they'd be shimmering microtones of gold: caramel, amber, spice. You can so easily fall into contemplating it. Oud invites admiration, and gets it.

I'm a double-naive about oud: first because I don't habitually wear perfume, and second because my culture doesn't prize this smell in the way others do. In Chicago where I live, you can't buy oud in every street market as you might throughout the Arabian Gulf, or find it smoldering expensively in a Chinese billionaire's home altar. I'm surprised to learn that tourists throng to Todaiji Buddhist temple in Nara, Japan, to visit a fabled chunk of eighth-century oud nicknamed Ranjatai. Tags on Ranjatai indicate where the emperor carved away slivers as gifts to various historical figures—for instance, in 1574 when Emperor Oogimachi awarded a tiny chunk to General Oda Nobunaga for his efforts in unifying Japan.[1] For me, oud is not a smell both elevated and deeply domestic, unifying the sublime and mundane. Unlike the Qatari men interviewed for the Al Jazeera documentary *Scent of Heaven*, I can't sniff Cambodian oud and leap, with a pang, to memories of my mother—or contrast this to Indian oud and recall my father.[2] For its adherents, the smell of oud runs deep.

Like many luxuries, oud is only one of this smell's names. It's also known as agarwood, aloewood, *gaharu*, *jinkoh*. Oud emerges in nature via happy accident. In a Southeast Asian forest, insects begin to feed on an evergreen tree of the genus *Aquilaria*. Some trees—less than 10 percent—will produce a dark, fragrant resin in an effort to contain the infection. The resin saturates the wood at the infection site, transforming living timber into a substance that resembles petrified stone: dark, hard, desiccated. This transformed material is oud. If the infection persists, the tree's resin continues to flow, slowly improving the oud's quality as it ages. At some point, when desire or greed finally overpowers patience, one cuts the chunk of oud from the tree and whittles any healthy wood away. The result is pure oud chips, which are packed loose into crates and sold to incense dealers in Southeast Asia, with particularly brisk trade throughout the Arabian Gulf.

The finest-quality oud is called "sinking-grade," so heavy with aromatic resin it sinks in water. The Japanese name for oud, *jinkoh*, simply means "heavy." Incense dealers can tell the quality of oud chips by listening to them fall through their fingers: oud chips should clatter audibly with a specific, dry music.[3]

Jinkoh played a starring role in *kōdō*, Japan's formal incense ceremony, which began during the Muromachi period (1336–1573), peaked in popularity during the Edo period (1603–1867), and waned after that.[4] A favorite pastime among the courtly and shogun classes, *kōdō* was essentially a memory game played with smells. Much as per-fumer trainees do today, beginning *kōdō* participants would start by identifying clearly divergent smells and advance to more subtle distinctions. *Kōdō* encompassed many classically beautiful scents—cloves, star anise, conch shells, musk, frankincense—but the undisputed star smell of the game was jinkoh. Contemplating jinkoh's different grades and scent qualities forms the core of many *kōdō* games. The Japanese call the highest-quality jinkoh *kyara,* a word that became synonymous in Japanese with first-rate beauty of any kind.

Kōdō asked its participants to "listen to incense" versus smelling it, a curious met-aphor until you realize that the Buddha's words were considered to perfume the air as he spoke. Bodhisattvas now listen for the Buddha's words by sniffing actual incense, attuning themselves quietly as they breathe the smells in.[5]

A distinctly literary ritual, *kōdō* encompassed hundreds of games each with their own rules. Here is a simple and delightful one you can try at home. First, the master of ceremonies hands out paper and pencils to participants. Then she distributes four scents wrapped in cloth and identifies each smell so participants can memorize each one. Let's imagine she chooses cinnamon, jinkoh, camphor, and cloves. She collects the smell samples again and then recites a famous four-lined poem:

> *I departed the capital*
> *Shrouded in springtime mists.*
> *An autumnal breeze blows here*
> *At Shirakawa Border train station.*

Next the emcee burns the four smells in turn, secretly scrambling their order and assigning one smell to each line of the poem. Here the emcee should display wit in her pairings. Perhaps she chooses to match camphor with line one of the poem; to her, de-partures seem chilly and anxious like this scent. She might pair the scent of cloves, which are actually flower buds, with the springtime mists line. She decides the autumnal breezes will smell of cinnamon, and jinkoh will smell of reaching one's destination, as represented by the Shirakawa Border station. The correct order of smells, as assigned by the emcee, is different from the original order of smells: camphor-cloves-cinnamon-jinkoh instead of cinnamon-jinkoh-camphor-cloves. Now it's up to participants to identify those smells in the order they were presented, while meditating on how each smell relates to its line in the poem.

Participants write down their guesses, and then a record-keeper quizzes them. Obviously, one hopes to identify all four smells, but even wrong answers are transformed into a story. If a participant identifies only one of the four smells, and that smell is jinkoh, the courtly explanation for this might be that he loves destinations more than journeys. If he identifies only cinnamon (autumn) and camphor (departure), the emcee concocts a different explanatory story: is he reminiscing about a bygone trip? Regretting a lost love in a forgotten city? Knowledge of scents is only one goal of *kōdō*. Its true purpose is to combine knowledge with a kind of aesthetic play: to spin a beautiful, witty, shared contemplation out of sniffing air.

Back in the modern-day world, civilization's consequences are rapidly catching up to oud. The best way to diagnose which infected *Aquilaria* trees are producing oud is to hack at the trees with machetes to check their insides. This practice has endangered many *Aquilaria* species such that wild oud harvesting is now strictly limited. Farmers can cultivate oud in plantations, but the process is maddeningly slow and the resulting scent disappointing. The disappearance of wild oud, combined with the slow arc of domestication, has created a perfect economic storm: oud's prices are both extravagant and always rising.

One can extract oud oil from wood to make perfume, or burn it as incense in the traditional Arabic way: Inside an incense burner light a piece of charcoal, then bury it in ash to insulate the heat. Place a sliver of oud over the hot ash until the wood smokes. Then enjoy the indirectly heated, luxuriant smell. Oud incense burners are handheld, so you can waft the oud's smoke inside your clothes and hair or pass the burner to a friend. A formal dinner in the Gulf often culminates with the host passing around different scents, including oud, for guests to perfume themselves. Guests arrive already perfumed and exit refreshed with new perfumes, layered onto the smells they wore into the meal.[6]

I sniff my wrists once more. Personally, I don't love the smell of oud, but I have come to respect it. Oud feels too baroque and civilized for me, but I'm intrigued by how its extreme civilization shades right back into wildness. Subversion is only sexy because there are rules.

CAMPHOR

It pierces the nose, spreading its frosty latticework throughout the face. After that first hit, the smell of camphor recedes into a pleasing, wintergreen roundness. Sniffing camphor is rousing, like swimming through a zephyr of chilled water in an otherwise warm lake.

Natural camphor comes from the wood of Southeast Asian laurel trees. Other wood resins, like frankincense and myrrh, are harvested by tapping live trees so the fragrant sap oozes out. Harvesting natural camphor requires chopping the tree down. You split the log to reveal fine crystalline veins of icy camphor. Passing steam vapor through camphorwood chips, then condensing the vapors, will capture the scent as a white, oily wax.[1]

Camphor is one of the rare substances that sublimates: from a solid state, it dissipates magically into the air without ever becoming liquid. The smell can repel insects, prevent rust, and fight microbes. With its strong trigeminal effect camphor's smell can reduce inflammation, numb aches and pains, clear the nasal passageways, and calm a cough. Chinese nicknames for camphor evoke its slightly otherworldly qualities: *ping-pien*, "ice flakes," and *lung-nao-hsiang*, "dragon's brain perfume."[2]

In medieval Europe, camphor was prized for its coldness and dryness, and as such considered more medicinal than perfume-like. "In contrast to aphrodisiacs, such as musks and other animal scents, camphor extinguished sexual excitement," as smell historian Jonathan Reinarz writes in his book *Past Scents*. Sanskrit poetry similarly extols camphor's iciness as a way to calm and soothe, associating camphor symbolically

EXERCISE

Navigate around a room by smell alone.

Clear a room of unnecessary clutter and furniture first.

Ask a friend to soak a few small sponges with the same essential-oil scent, then hide the sponges in spots around the room. Blindfolded, you smell the entire room, moving around it as you wish.

Then, spin yourself dizzy in another room (again, a friend can help minimize damage and injury). Return to the smell-infused room and find your way to each sponge, using only your nose's powers. (You're replicating another UC Berkeley smell study, testing humans' abilities to navigate a space via smell.[*])

Make sure to vary how intensely you sniff. Move your head around to get the full stereoscopic effect of both nostrils.

Repeat the exercise, this time with your friend as the smell navigator.[†]

---

[*] Ethan Walker, "UC Berkeley research shows that smell plays part in human navigation," *The Daily Californian*, June 26, 2015, https://www.dailycal.org/2015/06/26/uc-berkeley-research -shows-that-smell-plays-part-in-human-navigation/.

[†] This exercise was pioneered by the practice of olfactory artist Maki Ueda in her series Olfactory Labyrinths. Work #2 in that series asks participants to navigate mazes constructed from boards of fragrant woods. You complete the maze by following a single scent that connects its beginning to its end.

with the moon. In the tenth-century poem *Saundaryalahari*, camphor flakes falling from the lips of the goddess Devi cool the searing heat of three smoldering cities.[3]

In 1908 the emperor of Japan attempted to monopolize natural camphor, producing a frenzied stampede among chemists to synthesize it in a lab. Gustaf Komppa cracked this nut and began industrial production shortly thereafter. He also coaxed camphor molecules from plants other than felled laurel trees—specifically from alpha-pinene, an aromatic compound released by plants including pines and rosemary.[4]

Camphor's chilled smell sharpens attention. It's so distinctive that mathematician Francis Galton included camphor as one of his smells in a thought experiment he called "smell arithmetic." "Arithmetic may be performed by the sole medium of imaginary smells, just as imaginary figures or sounds," Galton wrote in a quirky 1894 paper. Using only trigeminal-forward scents like aniseed, ammonia, and camphor, Galton writes: "I taught myself to associate two whiffs of peppermint with one whiff of camphor; three of peppermint with one of carbolic acid, and so on . . ." Galton declared his experiment successful, adding and subtracting scents with giddy abandon, although he "did not attempt multiplication by smell."[5]

Burning camphor is used in Hindu, Buddhist, and Jain temples to activate the third eye, stoke intention for prayer, and purify the mind.[6] (Unsurprising, given that the etymology of the word *smell* in most languages is often intertwined with the word for "smoke.") Smells like camphor can demarcate a sacred place and time, clear a space for deep contemplation, then waft invisibly off. In its fragile intensity, camphor reveals how a smell can become a pop-up mosque or chapel.

FRANKINCENSE

It seems to shimmer with smell-notes like golden filaments. Because you heat the incense indirectly on top of a glowing charcoal, frankincense's smoke is mostly mundane and familiar—until it flashes with momentary brilliance once again.

This is a dense smell with opaque smoke to match: sweet, a bit cloying, more ambrosial than actually edible. It's vaulting and almost unreal. The scent resembles absolutely nothing from humdrum daily life. Insistently, even floridly, the smell of heated frankincense blocks out the everyday and suggests a world of only rarity and miracles.

Because I know about it only from the Bible—specifically the Magi's gifts to the baby Jesus—frankincense seems unearthly, potentially unknowable. Before I began this chapter I wondered: what *is* frankincense, anyway? Does it still exist? What makes frankincense on par with gold as a fitting gift to a newborn god?

IT DOES INDEED EXIST—you can buy a plastic baggie full of frankincense for ten bucks on Amazon. It's an aromatic gum resin from trees in the genus *Boswellia*, which grow in the hot, arid regions stretching from eastern Africa into southern Arabia to northwestern India. Do a Google image search on *Boswellia sacra*, the species that yields the finest-quality frankincense, and you'll find photos of a dense, scrubby plant, more bush than tree, slanting off against sand dunes or sheer cliff faces, as dashing and noble as an Arabian prince surveying his kingdom.

To harvest frankincense, one makes a deep incision lengthwise into the tree's papery bark and pale-golden tears of resin leak out. The resin hardens as it dries in the desert sun, and two weeks later it's collected.[1] Ancient traders loaded resin crystals into sacks and carried them from the remote desert into populated areas. Frankincense is as distinctly Arabian as the iconic single-humped camel. (These animals got domesticated in 900 BCE partly to haul frankincense more efficiently.[2])

Frankincense is synonymous with the entire category of incense—the word *frankincense* simply means "high quality incense" in Middle English. (Its other name, olibanum, comes from the Hebrew *lĕḇōnāh* or "milk," referring to the resin's milky-white color when fresh and liquid.)

Back at the dawn of the world, good smells like frankincense weren't categorized as neatly or rigidly as they are today. Good smells used to shape-shift in their forms and applications. You might rub them onto your skin or hair to smell nice. (And most early perfumes came as solids—specifically, waxy unguents—not liquids.) You could dissolve them into drinks or spice your food with them. You might ingest the smells for medicinal purposes. And you could burn them to create a pleasant ambience or worship the gods or communicate with the dead.

They smell of an apple bitten by a tender lass; of the scent of Corycian saffron; of a vine blossoming with the first bunches of white grapes; of fragrant sheep-cropped grass; of the myrtle and the Arabian harvester; of chafed amber; of the redolent pale flame from eastern incense; of the earth after a light sprinkle of summer rain; of a garland on tresses dripping with spikenard: all these fragrances, Diadumenus, cruel boy, are in your kisses. What would they smell of, if you gave them without stint or disdain.

—Martial, Roman poet

Of all these usages, however, good smells existed first—and often primarily—as heavenly smoke. The word *perfume* comes from the Latin *per fūmāre*, "to permeate with smoke" just as *incense* comes from the Latin *incendere*, "to burn."[3] Good smells often found their fullest expressions as incense, of which frankincense is the oldest and most venerable kind.

It's almost impossible to capture all the meanings and usages of incense across history and spiritual practices. But it's easy to imagine how incense may've been originally discovered—a happy accident of fire—and how well-suited it is for religious rituals of many kinds. Incense's smoke travels upward, connecting the everyday world with the heavens. The scent of incense visibly fills a room with an otherworldly, emotionally gripping presence. Incense literally gets inside you, curling up all the worshipers' nostrils together. Incense is where we get the very idea of "inspiration" from.

Frankincense was sacred but also mundane. Ancient Egyptian women used frankincense ashes as kohl eyeliner. When eaten, frankincense can be antibacterial and anti-inflammatory; it also fights indigestion. Then and now, you can burn frankincense to freshen your home, rub its essence into your beard or hair, perfume your hands and feet with it, or chew it to sweeten your breath. Burning frankincense gives off a particular molecule, incensole acetate, that's been found to alleviate depression and anxiety in mice. Frankincense might elevate one's spirits in ways beyond the religious.[4]

In smelling frankincense, I'd expected to contemplate the distant dotted line between Magi times—to me, a time associated with myth—and the reality of now. But so many things conspired to close and complicate that gap. Not only did I buy my frankincense on Amazon, I also bought my tiny incense-burning pot and the charcoal nubs recommended by their algorithm (supposedly perfect for hookah pipes). There's clearly a mini-economy of consumers buying incense and related accoutrements.

Even frankincense's actual smell, when I finally burned my own, pinged a distant but clear memory. I knew it might smell "churchy" in a way I could find repellent. I'm a recovering Roman Catholic with no lingering affection for organized religion. Yet it's very different to burn frankincense now—as an adult in my own home, to satisfy an intellectual curiosity—versus my childhood days when the smell felt foisted upon me during interminable Masses. Context matters—and it changes even ancient smells. Sniffing frankincense now makes me recall wild, beautiful illogical rituals I participated in at church as a kid, many of which foregrounded smells. On Saint Blaise's feast day in February, kids would kneel to have their throats blessed by waxy candles bound into a cross by red ribbon. I recall our candles had been slicked with some sweet-smelling oil, leaving a wet trace under our ears.[5] On another occasion in elementary school, exactly which one I've forgotten, we marched around the church grounds singing hymns at dusk and carrying a lit beeswax candle adorned with a paper skirt. We trailed a priest who swung a heavy, clanking incense pot and released wafts of frankincense that you'd catch only while turning corners in the gathering dark. "Smells and bells" religion can be a training ground for aesthetics, for seeking out a beauty that's both sublime and eminently available, beauty that's not fully explained by the world's limited terms.

Let's return full circle to my first question: why did the Magi bring Jesus gold, frankincense, and myrrh? Short answer: all three were equally valuable, so these were gifts that conferred profound respect. Longer answer: they were chosen symbolically by later writers to foreshadow Jesus's story to come. Gold symbolized Jesus's royalty, frankincense his divinity, and myrrh—a bitter aromatic resin we'll learn about next—his humanity.[6]

MYRRH

At first, myrrh smells identical to frankincense—but it takes only seconds for contrasts to emerge. Myrrh smells more earthbound. It lacks frankincense's soaring top notes, or rather it thins them out. Myrrh's smell is simpler, less frilly, more level-headed, a touch moody. If frankincense is champagne, then myrrh is bourbon or rye. I'm tempted to say its smell is realer than frankincense's, less alienated from the everyday world. Its beauty feels modest.

If the colors frankincense suggests are shimmering golden tones, effervescent and kinetic, myrrh's fragrance seems browner, slower, more matte. Sniffing myrrh calls to mind wet wood, rain-slicked mushrooms growing on a mossy log, caramelizing onions, brandy-soaked raisins baking, the bruised edges of overblown lilies. The fat middle note of the scent is astringent, drying out the nostrils pleasantly.

I smelled my myrrh two ways: first by burning it as incense (a tricky operation on a humid day) and also as tea, dissolving brown resin chunks—called "tears"—in boiling water. As tea, myrrh filled the mug with a dense caramel-colored liquid, heavy with particles like coffee dregs. While it was melting I could see golden fat bubbles staining the liquid's surface. For comparison I made frankincense tea, too. (Both of my resins were food-grade.) Clear amber tears dissolved into a chalky white liquid that flattened out the upper register of smells I'd detected in frankincense's smoke.

Myrrh is mentioned so often in the same breath with frankincense, I wasn't sure how closely related their smells would be. With the benefit of experience, I'd call them cousins or even doppelgangers. The two smells' histories are intertwined, but they diverge in usages and qualities. To anthropomorphize: if frankincense is the blond cheerleader, excitable and unapproachably perfect, then myrrh is her wised-up Goth friend, the darker and more turbulent horse.

Let's start with their similarities and shared roots. Myrrh is another aromatic resin from trees in the genus *Commiphora,* another scrublike tree growing more or less in the same geographic area as frankincense—the shallow, rocky soils of Ethiopia, Kenya, Oman, Saudi Arabia, and Somalia. Like frankincense, to harvest myrrh you cut an incision into the tree bark and let the resin ooze out until it dries into crystals. Dried myrrh resin is reddish and darker, less golden and refined-looking than frankincense. Like diamonds, myrrh and frankincense resins can be graded in quality by their clarity and color.

The taste of myrrh is extremely bitter: its very name stems from the Semitic word *murr*, meaning "bitter." Romans and Greeks used to steep sweet wine with myrrh to sharpen its taste and add balance and complexity to the drink. The wine given to Jesus hanging on the cross was likely mixed with myrrh.

Myrrh's astringency suggested to the ancients many medicinal uses. Many are now disproven but not all: myrrh may indeed reduce inflammation and kill bacteria.[1] Greek soldiers carried myrrh in their packs as an antiseptic for wounds.[2] Like frankincense, you could fumigate a room with myrrh, freshen your breath or hair with it, or even melt some to fix a broken piece of pottery.

Myrrh seems to shape-shift more than frankincense. Like frankincense, you can burn solid myrrh crystals as incense. But unlike frankincense, myrrh also exists as a liquid perfume. Consider this juicy quote from the Bible's Song of Songs: "My beloved thrust his hand through the latch-opening; my heart began to pound for him. I arose to open for my beloved, and my hands dripped with myrrh, my fingers with flowing myrrh, on the handles of the bolt."[3] The Songs writer may've been referring to stacte, an essential oil of myrrh that oozed naturally from the living tree without the need for incisions. You could also extract stacte from crystallized myrrh resin by heating it or dissolving it in hot water with balanos oil to create a semisolid unguent.[4]

Myrrh is both sexy and funereal: after all, can't a little death bring a little thrill and vice versa? The ancient Egyptians burned it at funerals and every evening to the sun god Ra at sunset, a practice most rigorously observed in the city of Heliopolis. They particularly loved myrrh for embalming. In the most lavish funerals, the Egyptians would empty the body entirely of its organs and fill the internal cavity with bruised myrrh, cassia, and other aromatics. Perhaps tellingly, burial spices almost never include frankincense.[5]

Myrrh is among the core ingredients of *kyphi*, a complex ancient Egyptian incense. Another shape-shifter, kyphi's recipe changes according to whatever historical source you consult. Dioscorides's recipe includes myrrh, honey, wine, raisins, juniper, pine resin, calamus (a kind of rattan palm), rushes (the same used in making the basket that rescued baby Moses), and a now-unknown plant called aspalathus. Egyptian recipes added other ingredients to kyphi: cassia and cinnamon, mint, henna, mimosa flowers, and mastic, another tree resin. You'd boil the ingredients and then roll them into incense balls to be indirectly heated. The order in which you prepared the ingredients apparently influenced the potency of the final result quite a bit. According to Plutarch, second-century BCE Egyptians burned frankincense in the morning, myrrh at noon, and kyphi in the evenings in their temples.[6] Kyphi may have perfumed the evening sunsets over the gleaming new pyramids, but the exact blend of smells drifting in and out are lost to us now.

I was already mentally personifying myrrh as a tempestuous woman when I stumbled on the Greek myth of Myrrha, also known as Smyrna. Ovid's *Metamorphoses* describes how Myrrha struggled against an unreasonable urge to commit incest with her father, which rendered her almost suicidal. Ultimately she tricked her father, King Cinyras, into sleeping with her in a debauch lasting several nights. When he discovered that he'd been sleeping with his own daughter, Cinyras tried to kill Myrrha but she narrowly escaped. She then roamed the earth pregnant and miserable for nine months, seeking refuge. Finally she asked the gods to take pity on her, which they did by transforming her into a myrrh tree. Shortly afterward, as a tree, she gave birth to the god Adonis. Her bitter tears—the myrrh's resin—distill this fable.

Myrrh brings with it dark, strong stuff: death, sex, unspeakable urges, wounds. Deep inside pyramids, it sweetens the inner cavities of dead Egyptians. It's the bitter undertow lurking in sweet wine. It bites back, resists the saccharine. It's the loose end that refuses tying. Mixed with finer aromatics in kyphi, it mingles with clouds of gnats ascending into the skies as the sunset burns and burns. Myrrh is propulsive, agitated, interested, and interesting. It smokes out bitterness and sends it aloft, as perfume.

 **FUNKY**

*Considering sexual attraction via smell, all the information conveyed
by bodily smells, and how smell can unwittingly become a tool of many
kinds of bias. This chapter also explores stinks we perversely love.*

SKIN

NEW CAR

CANNABIS

CASH

GASOLINE

MUSK

SKIN

I cannot describe how my lover smells, but I can describe how his scent registers inside me. It leapfrogs questions of resemblance or vocalization and smells simply like the very best emotions. I feel encircled, known. It smells like my blood pressure plummeting and stress hormones evaporating. (Science backs me up: your lover smells this way to you, too.[1]) A lover's skin smells like relief, like a concentrated jet of grace.

The scent of skin consists of three intermingling levels. At the surface level, we fight stinky BO with deodorants, showering, and fragrances. The middle level stems from cultural factors: diet, environment, and social norms.[2] Below that level—beneath the scented layers of sweat, lotions, and last night's meal—you'll find a person's baseline smell. This smell is unique, trackable to you as an individual, and telegraphs all kinds of information about your health.

LET'S START AT THIS DEEPEST LAYER AND MOVE UPWARD. Your body's baseline smell emanates from the entire body but its dominant note lifts from the skin, the body's biggest organ. Unlike synthetic fragrances designed to blare aggressively outward, this scent is quiet, amplified only by body heat. To observe it at all requires drawing close to someone. What secrets about a person does this scent disclose?

Meet the major histocompatibility complex, or MHC, a cluster of fifty genes that code for your body's immune system. "MHC genes are the most variable in all of nature," writes smell researcher Rachel Herz in her book *The Scent of Desire*. "Everyone, unless you have an identical twin, has a unique set of MHC genes." Your particular MHC genes form the underlying *genotype* of your immune system. The *phenotype*—or how those genes express themselves outwardly to the world—is emitted as your body's baseline scent. Because MHC genes are codominant, each parent's gene contributes its bit toward their offspring's immune systems. We choose our mate, among other reasons, because their bodies smell good to us—partly because their MHC genes, and the immune systems they express, differ robustly from our own.[3] Modern epigenetics thus bears out the Elizabethan practice of making "love-apples": a girl slips a peeled apple under her armpit, dances vigorously all night, then proffers the fruit to a lover, who lustily smells and eats it.[4]

The scent of these arm-pits is aroma finer than prayer . . .

—Walt Whitman, *Leaves of Grass*

Of course, mating by smell is as complex as humans are. Taking an oral contraceptive can scramble a woman's smell preferences. With their bodies chemically tricked into believing they're pregnant, these women prefer the smell of partners whose MHC profiles *resemble* their own—partners who smell like their blood-relations. Imagine the haywire of going off the pill, and simultaneously going off your partner's scent. How much is smell implicated in divorce?[5]

Mating isn't exclusively about procreation, of course. Although the subject is less well researched, smell plays a role in non-hetero mating, too. When asked to sniff

T-shirts worn by gay or straight men, gay men can identify—and prefer—the bodily smell of other gay men.[6]

Are these pheromones? Not exactly. The two biochemists who first identified pheromones in butterflies in 1959 describe them as "substances which are secreted to the outside by an individual and received by a second individual of the same species, in which they release a specific reaction, for example a definite behavior or developmental process."

A pheromone functions between two animals much as a military telegram should: it's terse, unsubtle, consisting of a blunt command that must be unquestioningly obeyed. These messages can be mostly distilled to fight, flight, food, or fuck. Insect pheromones telegraph danger to other members of their group, demarcate that group's territory, indicate an exciting new food source, induce a caterpillar to start its cocoon, or initiate feverish sex between two mayflies—always in a pretty automated fashion. One doesn't quibble with, or even consciously interpret, the nature of a pheromone's message. Pheromones insist on chain of command.[7]

Pheromones can be transmitted many ways. In most mammals they travel via the vomeronasal organ, or VNO. The VNO operates like a kind of parallel nose. At its front end, it consists of two tubes embedded in the hard palate at the roof of the mouth. Running parallel to the nose's olfactory system, these tubes send pheromones to an accessory olfactory bulb, which plugs directly into the brain's amygdala and hypothalamus areas. When an animal detects a whiff of pheromone, it sucks more air into its mouth to capture the pheromone fully. (You know that mildly humorous face horses make, flaring their front lip as if laughing? That's called *flehmen*, and it's the horse's way of sucking a pheromone deeper into its VNO. One way you could induce Mister Ed to "talk" would be to parade a female horse in heat offstage.[8]) The amygdala receives the pheromone's message and blammo! Mating ensues.

"In the majority of mammals, including those primates that have it, the VNO acts as a sex smell detector, passing its nerve impulses directly to the part of the brain that controls sexual behavior," writes evolutionary biologist Michael Stoddart in his book

Adam's Nose and the Making of Humankind. "The fact that humans don't have function-ing VNOs doesn't mean we can't, and don't, respond to human scents," writes Stoddart. "It means only that we lack the instinctive behavioral responses the VNO . . . facilitates . . . We remain in rational control of our responses."[9]

The baseline smell of your body is unique to you, much more so than a thumbprint. Anyone capturing or analyzing your baseline smell could, given a smell-matching da-tabase, identify you as an individual. (The East German secret Stasi police dreamed of such a smell-controlled future. They snipped fabric samples from chair seats used in interrogation rooms, building a database of dissidents' bodily smells.) Trained dogs hunt wanted people via bodily smell, and electronic noses at TSA checkpoints could theoretically do the same.[10] All they lack is a comprehensive smell database and the legal right to do this. You can't stop emitting your baseline smell willfully. If facial iden-tification technologies trouble you, smell identification should worry you more. How might privacy suffer if authorities eventually link your smell to you?

Fluctuations in your baseline scent can signal that you're getting sick. In one study, participants were injected with lipopolysaccharide, a bacterial toxin that provokes a swift, strong immune response. (The control group was injected with salt water.) Four hours later, researchers collected both groups' T-shirts, sliced out the armpit areas, and put these fabric scraps into bottles. Researchers asked other participants to rate the fabric scraps' smells for pleasantness, intensity, and healthiness. The sweat from toxin-injected folks was deemed more intense, more unpleasant, and less healthy-smelling.[11]

Many diseases announce themselves as smells. With the right training, a doctor (or a dog) can detect the distinctive smells of Parkinson's Disease, malaria, multiple scle-rosis, and cancers ranging from melanomas to breast and lung cancers.[12] Traditional Chinese medicine practitioners routinely smell their patients' bodies. For instance, an overly sweet smell like rotting flowers relates to the spleen meridian, a bodily area as-sociated with how well we manage thoughts and intentions.[13]

Inside our bodies, invisible processes whir—and these leak into the air as smells.

The slang term *funky* in Black communities originally referred to strong body odor and not to *funk*, meaning fear or panic. The Black nuance seems to derive from the Ki-Kongo *lu-fuki*, "bad body odor" and is perhaps reinforced by contact with *fumet*, "aroma of food and wine" in French Louisiana. But the Ki-Kongo word is closer to the jazz word *funky* in form and meaning, as both jazzmen and Bakongo use *funky* and *lu-fuki* to praise persons for the integrity of their art, for having "worked out" to achieve their aims. . . . For in Kongo the smell of a hardworking elder carries luck. This Kongo sign of exertion is identified with the positive energy of a person.

—Robert Farris Thompson, *Flash of the Spirit: African and Afro-American Art and Philosophy*

NOW LET'S CONSIDER THE MIDDLE LAYER of our skin's complex smell—that whiff of each person's smell that betrays their culture and informs social interactions between groups. The fact that outsiders supposedly "reek" explains, and falsely justifies, a lot of xenophobia throughout history.

Writing in the fifth century CE, classic Sanskrit poet and dramatist Kalidasa remarked: "Every man has confidence in those of the same smell." It's an observation borne out by the many cultures—Maoris, Eskimos, Arabs, Indians, and others—who greet a person by first sniffing her scent. A person's smell tells you a lot instantly about their diet, occupation, living environment, and hobbies—and to what degree all of those things resemble, or differ from, those of the sniffer.

George Orwell once observed, "The lower classes smell," which summarizes most people's reaction to anyone who smells unfamiliar. To smell different from oneself is, generally speaking, to smell bad, period. Writing around the same time as Kalidasa, the Gallo-Roman bishop Apollinaris Sidonious wrote a letter home lamenting the stink of foreigners as he traveled:

> Ah, poesie is harde, cast as I am amongst hirsute hordes, defened by the stryfe of the German tonge, forced to laud the songes of a noisome Borgundian with rancid fatte in his hair when all I feel is disgust. Happy thine eyes, happy thine ears, and even thy nose . . . for it is not forced to sniffe the stench of garlic or onions ten times each morn.[14]

"All the persecuted stink," wrote one English author in 1830. "One of the first re- ceipts [recipes] for having a man persecuted, is to impugn the credit of his corporal presence." And majorities have been impugning minority groups for ages using their noses. Since the Middle Ages Jews have supposedly reeked of *foeter Judaicus*, a goat-like stench akin to the devil's scent. For a time, Christian baptism was deemed to eradicate this smell. But the supposedly foul scent clung to the poor, overcrowded shtetels where Jews lived and later still was considered innate to Jewish bodies, a quasi-genetic form of racial identity. The Nazis seized on this stereotype and elevated it to ideological gospel. "The Jews [have] a different smell," claimed Hitler, rendering them permanent "strangers" in his Reich. In a 1939 public speech he elaborated: "Racial instincts protected the people; the odor of that race deterred Gentiles from marrying Jews." Olfactory discrimination played a small but not insignificant role in the Nazi genocide.

Black Americans have been similarly tarred by whites on the basis of smell. "As early as the eighteenth century," writes historian Mark Smith in an article about olfactory stereotyping, "efforts to explain why black slaves . . . supposedly possessed a distinctive odor flitted between environmental explanations (they came from hot climates, they did not wash, their food was different) and quasi-genetic ones (their skin simply emitted a rank, fetid odor . . .)." Even a famed anti-slavery supporter, physician Dr. Benjamin Rush, cited Africans' vulnerability to leprosy as a fundamental cause of their skin's blackness, "rendering their skin 'black, thick, greasy' and causing it to 'exhale perpetually a peculiar and disagreeable smell, which I can compare to nothing but the smell of a mortified limb,' " in Smith's words. Bodily smell even played a decisive role in the 1896 landmark Supreme Court case *Plessy v. Ferguson*, in which a light-skinned Black man, Homer Plessy, dared to sit in the whites-only carriage of a train. Louisiana's prosecuting attorney, Louis H. Ferguson, argued that the evidence of the conductor's nose, not his eyes, identified Plessy conclusively as Black. Case won by the prosecution.[15] In all cases, the difference in bodily smells between races serves to justify their separation, and rarely are both smells considered equally pleasant.

However many ways this history stinks, the story of socio-ethnic smell isn't uniformly awful. Bodily smells sometimes promote diplomacy between social groups.

Kate Fox, codirector of the Social Issues Research Centre and author of *The Smell Report,* describes the role smell plays in certain aboriginal cultures in coordinating social mixing and marriage. The Amazonian Desana people believe each tribe carries a specific odor, and marriage must unite people of different odors. To celebrate a new union, Desana families exchange ritual gifts, favoring meats and differently scented ants. The Batek Negrito people of the Malay Peninsula also marry across odor lines, and even prohibit sitting too close to someone of the same odor. It's thought that prolonged mixing of too-similar smells breeds disease. The Temiar people, also of the Malay Peninsula, are mindful of smell-mixing to an unusual degree. They believe each person's "odor-soul" is located in the lower back and vulnerable to disturbance. If you must pass behind someone's back closely, a Temiar calls out "odor, odor" to excuse the personal intrusion.

Perhaps the most nuanced social users of smell are the Ongee tribe of the Andaman Islands in the Indian Ocean. To refer to himself, an Ongee points to the tip of his own nose: *smeller, c'est moi.* Their customary greeting begins by asking: "*Konyune onorangetanka?*" or "How is your nose?" If the person replies that they are "heavy with odor," it's polite for the greeter to inhale, absorbing some of the person's surplus scent. Correspondingly, if the greeted person feels light on odor-energy the greeter exhales, blowing additional scent in their direction. In the Ongee worldview, bodily smell isn't good or bad. It's simply energy to be exchanged and kept in balance.[16]

Many forms of greetings bring us close enough to sniff a person's body—that's often the point. Cheek-kissers may not always inhale each other's scent, but the world's nose-kissers generally do. Across the Gulf region and parts of the Arab world, men "nose-kiss" on a regular basis. For a greeting between men of different social classes, the younger or lower-status man will kiss the senior man's nose. Social equals will touch, rub, or press their noses together. Two Arab brothers might clash noses gladiatorially in happy greeting.[17] We sniff each other, at first cautiously and then with growing enthusiasm. Nose-kissing is a rapprochement with noses.

THE SURFACE LEVEL OF OUR SKIN'S SMELL is both the least interesting and the most profound. It can reveal our racing thoughts, or the heated pendula of exertion, weather, and mood.

On a 100-degree day your blood vessels dilate, bringing your hot blood closer to the skin's surface. Blood plasma evaporates through the skin, becoming odorless sweat produced by the eccrine glands. Sweat's evaporation reduces your body's temperature while leaving your blood denser, thicker, more logy. Drinking ample water liquifies your blood plasma anew.

It's your apocrine glands that produce the stench of BO. Activated at puberty, these glands concentrate in the hands, cheeks, breast areolas, and wherever there's body hair: the scalp, armpits, groin. They're at work in the secret, tangled spots of the body where sweat can't evaporate easily from the skin. The apocrine glands release fats and proteins tinged by what one eats. Skin bacteria are drawn to feed on these fats and proteins, yielding stinky chemical compounds. Aprocrine glands also work overtime during stressful situations, which gives emotional sweating a more pungent scent than simple heat-reduction sweating.

Deodorants either kill or deactivate these skin bacteria and their noxious activity. Antiperspirants work differently: they block sweat glands in a particular area so they don't produce sweat at all. If you block your armpit glands, you're only re-distributing sweat production to other glands throughout the body—but that might be desirable if those glands appear on better-aerated stretches of the body than the armpit.[18] Futuristic new deodorants may target only the skin bacteria responsible for the worst-smelling compounds.[19]

The body's surface-level scent reveals more than physical exertion or overheating. Clinical trials reveal all the emotions we can detect by smell alone: joy, fear, frustration, sadness.[20] We can smell these emotions in strangers, even more accurately in a loved one. Our skin is constantly radiating information into the air, information that ranges from epigenetic to momentary: about our immune systems and health, diet, and culture, even the thoughts flashing inside our heads.

I'VE BEEN INHALING MY PARTNER'S SCENT for a quarter century. It's corny but accurate to say his smell is an olfactory smell-track of my adulthood (and mine of his). Burying my face in his neck is a *mise-en-abîme* with thousands—maybe millions—of repetitions. All those ghostly embraces meet, are pinned to one another like tissue-paper dolls in a thick stack, focused on his smell. Inside the crook of his neck, amid the prickling hairs and warmth, I breathe him in again. How to describe this scent? For me, it stops the clocks. It clicks with deep sense. It's capacious, witty, the deepest and most beautiful private joke. (I've been sniffing his T-shirt for twenty minutes to help me write this. Life is ridiculous.) Have I been tunneling for years toward this smell, or do I now radiate myself outward from it? Who cares anymore?

NEW CAR

First comes the airlock as the door clicks shut. Even if someone is sitting in the passenger seat right next to you, the sense of privacy is total. This air, this tight enclosure of space, is all yours.

The smell inside a new car is chemically volatile, mildly unhealthy in a satisfying way. There are dashes of hot plastic, leather, industrial carpeting, rubber. The air feels thick with particles; always it seems a bit too warm inside. It's an *alien* smell, and this aspect is crucial to its charm. One pictures robotic arms busily wielding hot-glue guns, nestling door handles into upholstery, fitting rubber flashing around each window. No human inhabitants have disturbed the atmosphere of this little planet.

Responses to this smell are vast and wider-angle than I'd supposed. Turns out new-car smell captures diverging ideas of capitalism in the two superpowers of our time and competing consumer fantasies of each.

Picture an American and a Chinese consumer, each strolling into a sparkling show-room on opposite sides of the planet. Each roves the floor a bit, circles a particular car, and then settles into its driver seat. The American inhales the new-car smell with abiding satisfaction. It's a recent whiff of the factory, a deeply priming one. It speaks of newness: crisply beveled edges, responsive suspension, a gleaming engine optimized for performance. To this theoretical American, it smells of technology and the future unfurling before her. A new car enlarges her physical person and amplifies her power. She can speed through vast landscapes whose very American details she already knows how to picture from TV ads: undulating wheat stalks, tumbleweeds, bleached-out road-side stands, hairpin turns through valley gorges amid ponderosa pines.

Sitting inside his new car, the average Chinese consumer inhales with very different expectations. He wants to smell total blankness, and the facts actually support his worldview. The new-car smell Americans find so intoxicating *is* indeed a teensy bit toxic: it's just brand-new plastics, leather, vinyl, and chrome releasing off-gases. The scent intensifies as temperatures rise and tails off rapidly as the gases are fully released. Chinese consumers are keenly sensitive to new-car smell and rank an odorless interior among their top priorities in a car purchase, up there with factors like engine performance, fuel consumption, and safety.[1]

This preference is so real, it fuels actual jobs. Ford's Chinese research labs employ human smellers—known internally as "Golden Noses"—to sniff every component of a new car's interior and ensure it smells of absolutely nothing.[2] Recently Ford took things a step further by filing a patent for a car-odor-removal process. Still theoretical and applicable only to semiautonomous or self-driving cars, the patent's technology consists of "baking" the interior smells out of a brand-new car. Right after the car rolls off the assembly line, it will drive itself off to a sunny outdoor spot, crack its windows slightly, and then turn the engine, heater, and fan on high. The patent's technology includes air-quality sensors so the car itself can determine when its interior is odorless enough to stop the process.

Why this difference in consumer tastes? Perhaps it's explained by the different realities the two buyers traversed to reach the showroom. The Chinese consumer may live in an olfactory-heavy environment with rampant air pollution and ever-increasing urban density. Factory newness may not hold the same cachet in a country bursting with factories.

What's more, buying a brand-new car in China is a mass-luxury event; only one in four Chinese owns a car. One auto expert described Chinese car buyers' idea of luxury as consisting of thoughtful layers. If you buy a luxury car and pull back the carpeting to reveal unadorned metal, that's considered low quality. Each layer of the car should look and feel consumer-ready, even—perhaps especially—the layers below the surface. To a consumer with this mindset, a car still reeking of factory newness might seem insufficiently polished.[3]

But the difference might be more than cultural. According to *New Scientist*, many Asians' bodies produce less of a particular enzyme that breaks down ethanol and acetaldehyde, a major VOC released by car interiors. Combine this heightened sensitivity with the high levels of air pollutants inside new cars—often ten times the regulated limit in China—and new-car smell may literally irritate Chinese consumers more.

New-car smell might someday vanish entirely, joining the dozens of ghost scents lost to history. A century from now, only an electronic nose might sniff its way through a car's interior, wrinkling in disgust at an offending smell before obliterating it with heat. This is sad for us Western sentimentalists fond of this funk. But it does clear the way for a new and more purist form of olfactory cachet: a new car that smells truly blank, poised to absorb the many smells its owner brings to it.

CANNABIS

The smell is big, bodacious, untidy. It's got a wide amplitude, saturated with information. The form of cannabis I'm smelling right now is called "flower" and is, literally, flowers: specifically, whole dried buds from a cannabis plant. Much like strict Catholics, pot growers tend to prize virgin females above all others. (Unfertilized female plants yield larger, plumper, more resinous flowers.)

This strain, Super Blue Dream, smells sour and incandescently green. Whiffs of chilled mint and pine resin waft above base notes of a sweet, tarry dankness. Dried flower's smell does resemble that of burning marijuana. But flower's scent is more delicate, with discernable layers.

It's also lively. Each sniff crackles with a slightly different combination of scents. Like a climbing vine, the smell moves vigorously and resists containment. Stored in a sealed baggy inside an airtight container, my weed still manages to telegraph whiffs of itself from across the room.

It's a polarizing smell, one that immediately demands to know your stance on it. Are you an insider, or an opponent? I'm unsure how to answer this question. My own position splits hairs: I'm theoretically pro-cannabis, yet practically inexperienced. The thicket of judgment and heated opinions attending this smell almost overpowers the scent itself.

Cannabis's smell has long existed in a gray area between categories: food, medicine, drug, religious incense. In this it resembles other commodities we now lump together as the early "spice" trade. Many organic materials—cinnamon, roses, opium, indigo—were once used interchangeably as food, medicine, perfume, even artists' materials and dyes. Pharmacists mixed up recipes for all of these and dispensed them to diverse buyers from the same shop.

Cannabis still occupies a highly contested gray area today: to its opponents, it's a dangerous "gateway drug;" to its consumers, it's a mostly harmless, pleasant habit; to legalizers, it's an untapped source of tax revenue. But the gray area swirls around the smellers as well as the smell itself. A friend once observed to me: you can't smell pot at all when you're the one smoking it. You only smell pot when you're outside of it. It's a trenchant remark that reveals a lot about how cannabis is wielded by police to determine who's inside versus outside, who's allowed to relax this way and who's not.

To a white person, smelling like pot is usually embarrassing in a nonserious way: a minor offense, a momentary indulgence. You might be able to swat it away as you'd aerate a smoke-filled room or car. An authority can decide to hassle you about it, but often they won't bother.

To a Black or brown person, smelling like pot can be similarly laden, but it can also become a much graver matter. To a racist cop sniffing it around a person of color, cannabis can smell like a provocation. A justification to investigate, to root out whatever other infractions they can find. It smells like the leading edge of an unstoppable process, a trail of scent that can deepen and complicate into the festering stink of mass incarceration.[1]

Cannabis's smell is undergoing a rapid cultural transformation in America. The state where I live, Illinois, legalized recreational cannabis use in January 2020, making it one of fifteen states offering recreational access to adults. As of this writing it's legal for medical use in thirty-four states as well.[2] Legalization should make marijuana use more mainstream, reshaping our associations over time with its scent. It should also, theoretically, stop providing a pretext for police harassment for Black and brown consumers—although that's not yet happening.[3] In fact, nationwide Black people are 3.6 times more likely than white people to be arrested for cannabis use, despite similar usage rates among these groups; that disparity has worsened with legalization in many places.[4] So cannabis is a complicated scent, for reasons both inherent to the plant and external to itself.

As a biological defense system, our sense of smell makes swift and often irreversible judgments. That makes cannabis rare for another reason: it's a smell I've encountered but haven't fully made up my mind about.

Smelling exhaled pot on the sidewalk captures the tail end of a cycle, and the least interesting bit at that. For me, getting acquainted with cannabis's smell brings with it the awkwardness of initiation. It means buying pot, smoking it, mastering lingo and techniques, the whole corny bit. Inviting my body to surprise me. Deciding if, and how, I'll fit into a group. Admitting my ignorance on several uncomfortable levels.

This situation calls for either adolescent fumbling or, I'm realizing, a humble but gonzo confidence. So be it. Enter the gawky forty-something white lady with her nostrils flared. She is slightly ridiculous, but she is game.

SEVERAL WEEKS AFTER I BOUGHT IT, my flower's smell has muted slightly. The loud whiffs that so readily escaped my storage container have retreated inside it.

It took me that long to realize that cannabis smells like plants. My flower is a recently alive plant that differs subtly in smell from strain to strain and whose scent changes over time. Drying cannabis preserves and extends the flower's shelf life, the same reason you might dry spices, fruit, or tea.

What I'm smelling in my weed is *terpenes*, aromatic compounds produced by all plants. I hadn't understood before that different strains of cannabis vary widely in their smells, and cannabis aficionados are also master-sniffers. Cannabis distributors use terpenes to verify the exact strains they're getting, which might go by dozens of different brand names. Patrick Matthews, author of *Cannabis Culture*, offers an explanation logical in the pre-legalization era: "People who take illegal drugs usually try to compensate for the lack of labelling, quality control or legal comeback on dodgy suppliers by becoming as well informed as possible."[5] What might once have been a sensory precaution is—in the legalized era and for privileged consumers like me—shading into a farm-to-table-type enthusiasm. Cannabis has always had specific smell-notes and terroir, just like coffee, cheese, or wines. It's just getting less ludicrous to talk about it that way.

How many strains are there? Certainly more than I realized. The two main types of cannabis, sativa and indica, branch out into many, many hybrid strains. Indica and sativa plants differ in appearance, psychoactive effects, and smell profile. Sativa plants are tall and loosely branched, known for providing an awakening, creativity-inducing high. As a group, sativa plants emit smells that are more fruity, peppery, often acrid—"cat pee" is a frequent comparison. Indica plants are less tall, more conical in shape with dense, compact branches. Indicas encourage sleep and reduce anxiety, with a range of scents to match that mood: earth-bound, umami-tinged, and deep.[6]

My strain, Super Blue Dream, is a sativa hybrid known for the terpenes alpha- and beta-pinenes, redolent of pine trees, and myrcene, which smells earthy. Myrcene is the most prevalent terpene across cannabis strains; mangos, hops, thyme, and citronella also smell of myrcene. A friend let me sniff the differences between my flower and hers, another sativa hybrid called Candy Land. Her cannabis was chillier in the nose, much more minty. It made mine smell sun-drenched and citrus-forward by comparison.

Terpenes influence cannabis's potency as a drug in an indirect, almost sneaky way. THC is the best-known cannabinoid, a category of chemicals interesting to researchers for their potential medicinal uses. (Another cannabinoid you've likely heard of is CBD, a non-high-inducing chemical that may promote wellness.) THC is the psychoactive chemical in cannabis that makes you high; by itself, it has no smell. Terpenes, the aromatic compounds in cannabis, don't produce any high—but terpenes do influence how THC gets absorbed by the body. An extra-stinky strain of weed is not necessarily more THC-rich or potent as a drug. But those stinky terpenes do influence how you will experience THC's psychoactive effects.[7]

If terpenes don't make you high, can you enjoy weed without them? Apparently not. Nobody takes Marinol, a THC-only pill prescribed to cancer patients, for fun. This odorless, terpene-free drug is reputed to deliver a jarring, unpleasantly medicinal high. Terpenes—that is, the smells of cannabis—soften and humanize the experience of consuming it.[8]

Each cannabis terpene smells distinctive, with its own potential superpowers. Limonene smells like citrus with research suggesting antidepressant and antibacterial effects; you can also smell limonene wafting from rosemary and juniper plants. Linalool smells sweet and floral, a scent cannabis shares with lavender. Linalool may fight convulsions and anxiety, with effects some scientists liken to popping a Valium.[9] Caryophyllene also scents black pepper, cloves, and cotton plants; this terpene helps treat autoimmune and gastrointestinal problems. Ocimene imparts a sweet, woody, herbal smell and may fight infection, bacteria, and viruses; mint, parsley, and orchids also emit ocimene.[10] Earthy-smelling myrcene can reduce inflammation.

Thinking about cannabis as a plant, on a continuum with food or medicine, prompts a question: Legislatively speaking, how did cannabis wind up in the heavy-drug category with opium and cocaine, and not in a milder category like coffee or alcohol? The whiffs of sanctimony—and answering defensiveness—clinging to the smell of cannabis can be explained by its history.

THE EARLIEST RECORDED MENTIONS of cannabis were medical. In the third millennium BCE, Emperor Shen Nung documented his knowledge of cannabis in the *Pen Ts'ao Ching*, which became the standard reference work for Chinese medicine. According to *Cannabis: A History* by Martin Booth, Shen "advised the use of 'hemp elixir'—probably a tea made of cannabis leaves and flowers—to treat a wide range of ailments from gout to malaria."[11] Later Chinese physicians found additional uses for the plant they called *ma*—for instance, mixing cannabis resin with wine or the herb aconite as an analgesic for surgery.

Cannabis traveled to the Indian subcontinent with nomadic tribes around the same time Shen began his experiments. Cannabis features prominently in the Vedas, four seminal Hindu texts from 1100 BCE; the god Shiva claims it for himself, earning the name Lord of Bhang for the milky-sweet tea of cannabis leaves still drunk in his honor at weddings and festivals today. A classic recipe of bhang smells of rose water, cardamom, anise, garam masala; drunk cold, the dank note of cannabis is subsumed in a larger, complex smell.[12] Even abstemious Hindu sects partake of bhang, as do many Buddhists. They considered cannabis a food and a mild drug supportive of religious ritual.

Aryan tribes spread from the Indian subcontinent to Persia, Asia Minor, Greece, the Balkans, eastern France, and Germany. Its scent clings to the plant as it travels: the word *cannabis* derives from Aramaic and Hebrew words meaning "fragrant cane." Early Arabic texts refer to the plant as "the blissful branches," "the shrub of emotion," and "the morsel of thought."

Cannabis grows like a literal weed everywhere. It's logical that people would find uses for every part of an abundant plant: rope, paper, fabric, teas. But what you can make from its innocuous leaves, stem, and seeds differ a lot from preparations using its THC-rich flowers and their resin. Concentrating the resin produces hashish, a highly potent drug that has cast a long shadow over all varieties of cannabis for centuries.

Nineteenth-century Europeans and Americans alike took cannabis in over-the-counter medicines to curb anxiety, quiet a cough, or dull pains. It came packaged in many forms, most of them odorless: pills, syrups, tinctures, snuff. But cannabis never worked well as an industrialized pharmaceutical. The medicine's strength varied widely, even between bottles of the same brand. Cannabis was slow to take effect, so impatient consumers took too much. Cannabis tinctures separated into layers, making overdosing easy. You can't kill yourself with a cannabis overdose, but you *can* scare the bejesus out of yourself with hallucinations—an alarming side effect for a mere headache remedy. It took knowledge to consume cannabis moderately.

Cannabis's reputation was also tainted by exoticism. Striding about their colonized lands, Westerners smelled cannabis constantly—but usually as part of indigenous rituals to which they weren't privy. Those Europeans who did experience cannabis directly were antiestablishment poets, artists, and intellectuals. Many of these expats liked hashish's aggressive high and skipped milder forms of cannabis entirely. As this group tried opium, then hashish, several adherents published lurid exposés of each drug, which sold like hotcakes to Western readers.

In the late 1800s, British authorities in India decided prohibiting cannabis within the colony was impractical given its widespread use in Indian religions, socializing, and cuisine. But tolerance without understanding inevitably curdles. The scent of cannabis became, to them, the stubborn stink of otherness. In India, South Africa, and many other colonized places, cannabis helped Europeans justify to themselves why they considered brown people lazy workers, unmastered by the clock, fit for subjugation. That dank cloud settled over cannabis and followed it.

By 1900 "an estimated 3 percent of the American population medicinally [were] addicted to opiates," writes Booth, creating in consequence "a climate of considerable caution. Drugs were, in short, deemed un-American." That caution shaded rapidly into xenophobia and found culprits in Chinese and Mexican immigrants as well as Blacks. In cracking down on opium, cocaine, and then hashish, smoking cannabis became tinged with racist suspicions.

Enter American Prohibition, which banned alcohol but did not address cannabis use—a loophole many Americans cheerfully exploited. They frequented "tea pads"—speakeasies for marijuana that also provided community to Southern Blacks transplanted to Northern cities in the Great Migration. Crowded into underground jazz clubs, Blacks (and the minority of whites who joined them) mixed freely, tapping their toes in a dense, aromatic cloud.

Then–Federal Bureau of Narcotics chief Harry Anslinger took aim at cannabis and won. Together with powerful allies—New York City mayor Fiorello LaGuardia and newspaper magnate William Randolph Hearst—he enacted a federal ban on the drug in 1937. Anslinger's genius was rebranding cannabis with a foreign-sounding name, "marijuana" (which he spelled "marihuana"). That exoticism made it easier to lambast Mexicans, Filipinos, Greeks, Blacks, Spaniards, Turks, and Latin Americans as "marihuana smokers" whom the drug supposedly made violent and sex-crazed. Did they reek of weed or simply of non-whiteness? To Anslinger and others like him, the two evils merged.

By the 1950s, according to Booth "cannabis, the black man's narcotic, was widely regarded as more dangerous than heroin or cocaine, not because of its potential for addiction but for its facilitation of multi-racial sexual communication." In the 1960s, white hippies signaled their distrust of the establishment—Western, capitalist, and white—by marching in civil rights protests with Blacks, dressing in Indian paisley-prints, burning patchouli incense, and puffing like crazy on "marihuana cigarettes." Generational and political differences were demarcated literally in smells.

As past shades into the present, cannabis' associations with Black and brown people have only gotten more durable. Why? Probably because the smell was useful to authorities. It became an instrument of institutionalized racism. It assisted directly in those communities' subjugation. Like other smells deemed foreign or "other," the smell of cannabis justified, even clinched, discriminatory actions that we were all too ready to perform. (For more on this theme, see the skin section, chapter 6.)

IT'S TIME TO COME FULL CIRCLE and finish what I started: from sniffing my flower to smoking it.

I unfurl a brown paper bag of its contents: my sealed container of cannabis, a hand-held grinder that crushes buds into powder, and a one-hitter pipe in marbled emerald green glass. I also own a vape pen, which heats up the flower without incinerating it. This is both less damaging to your lungs than inhaling smoke, with the side benefit of reducing the incriminating smell to almost nil. (If you've ever wondered why so many Millennials and Gen Z types vape, they aren't all switching to a more techno-forward form of nicotine. They're also vaping cannabis in a more discreet, normalized way.)

My dried flower seems to wake up in the nose as I prepare it: grinding it releases a bloom of scent, lighting it on fire yet another. The smell returns: big and sour, incandescently green, minty, tarry, high-res again with details.

I might be ready to enjoy smoking cannabis, but I also can't unknow what I've come to learn about this smell. I can't separate out the stink of racism or the reek of my own privilege. There's no resolving any of those paradoxes, either. It's all bound up in the same complicated scent.

EXERCISE

Bring more curiosity to stinks.

Prepare yourself by carrying around a jar of coffee beans or a tube of VapoRub. You can quickly sniff the former to erase unpleasant smells from your nose. If that's not sufficient, rubbing VapoRub over your upper lip should do the trick. If your bag is already bulging with stuff, simply avail yourself of nature's smell-eraser: the crook of your elbow. Sniffing your own skin will clear your olfactory receptors and reset your nose.

The next time you encounter a true stink, control your urge to recoil and observe it for a few beats longer. Sure, it's gross—but smelling it probably won't hurt you. What else can you notice in the smell besides stench? Inhale, exhale, observe. Then reach for the coffee beans.

I imagine something large and shameful catching fire and going up in smoke, like a burnt offering or a historical monument in smell. I take one hit, wait, and then let the rest burn.

CASH

Paper money—by which I mean U.S. dollars—smell faintly vegetal and musty. It's a smell that barely registers in the nose. Sniffing a dollar bill deeply, I thought of an un-air-conditioned library, the peculiar stillness of a warm room crammed with paper. I tried steaming a crisp one-dollar bill to coax more smells out of it. Besides setting off our overly sensitive fire alarm, the experiment's olfactory yield was puny. The warm, humid air brought out a slightly smoked note, like the smell friction brings out in ropes.

Scruffier bills smell more: less of the clean, desiccated top notes of a library, and more of a greasier middle note from the street. The smell makes you recall musty basements, locker rooms, the well-rubbed insides of old leather work gloves.

Sniffing a much-used dollar bill cues up a cinematic sequence, almost automatically, behind your closed eyelids. You picture the many environments these banknotes have traveled through since their creation: Printing-press vaults. Steel-encased trucks. Unseen black interiors of ATMs. Pockets, wallets, cash-register drawers, counting tables next to neatly packaged bricks of cocaine. Briefcases, birthday cards, G-strings, laundry dryers, piggy banks, tip jars, under the pillows of gap-toothed children. All through their journeys, banknotes absorb smells indelibly and carry them forward on their backs.

Traditional "paper" banknotes aren't made of paper at all, but fabric. The preferred blend is cotton and linen, usually heavier on cotton because it's more durable; the exact proportions vary from currency to currency. A brand-new banknote derives most of its faint smell from the inks printed on it. Each currency's ink recipe is a jealously guarded secret by its central bank and their currency-printing partners. So the weak scent of fresh paper money is a black box; we cannot know exactly what we're smelling.[1]

But paper banknotes really acquire their scent with use: via the secret oils of many, many hands and all the microbes our hands carry. (Incidentally, the same is true of spare change. Metal coins don't smell like anything when they're cold. When you touch a coin made of iron, perspiration in your skin causes the iron atoms to gain two electrons. These iron atoms further react with your skin's oils, yielding volatile compounds including 1-octen-3-one, that dank coin smell. A similar process happens with copper coins.) Cash smells like you and me, like our hand-to-hand contact whether friendly, hostile, or neutral.[2]

I'd assumed it would be easy to find a wide range of banknotes to sniff, from clean bills to dirtier ones. But when I opened my wallet, I only found clean bills. I asked my husband and a few neighbors, offering to trade a crisp bill for a scruffy one. Surprisingly, they also only had clean bills. I widened the net and asked more friends and neighbors; same result. One friend I asked didn't have any scruffy bills, but she did have a theory explaining why: our neighborhoods are too tony for scruffy bills. Rich communities, she argued, have cleaner cash.

My friend has had a lot of personal experience developing this theory. Her immigrant father first noticed this working a series of cash-heavy jobs in Chicago's downtrodden West Side. Working first as a cashier for a wholesale trading company, later at a dry cleaner's, he handled cash sourced from neighborhoods across the city. The skin of his hands was always irritated by nonstop contact with chemicals and cash. Doing his register counts, my friend's father could fan out the bills and recognize the crisp celadon cash of the Gold Coast and Lincoln Park from the oily, well-crumpled dollars from his West Side neighbors. Rich people enjoy so many thoughtless perks, he'd remark later to his family. Including fresher cash.

My friend later became an artist, supporting herself with service jobs where she dealt with a lot of cash. For a while she was a barista in an Uptown coffee shop across from a halfway house and a wide swath of Section 8 housing. Like her father, she could tell the difference between stinkier, crumpled cash from across the way versus the influx of stiff, deodorized cash that arrived with the office rush hour. But the grubbier cash pouring into her register also came tinged with an odd kind of friendship. She recalls several down-and-out types who parlayed one dollar spent on coffee into a full afternoon's access to indoor heating. These folks stood sentry over the coffee shop, shooing away more dangerous types, swept up and cleaned bathrooms for her, helped her open and close the store. Their dollar bills came dredged up from sodden pants, sweat-soaked, smudged with cigarette ash, but also plaintive and human.

My friend's anecdotes are just stories, of course; they don't prove any grander scheme at work. And yet it darkly illuminates this scent. I imagine banknotes enter a given community, perhaps via ATMs, and those notes just keep on circulating locally. It's logical: wealthier people use cash sparingly, usually only for minor expenses. Cash might leap from a rich community into a poorer one—say, by dropping pennies into a fountain, feeding a parking meter downtown, dropping a single into a busker's guitar case. But it rarely moves in the opposite direction. Poor folks don't spend their cash in rich neighborhoods if they can help it. Cash, particularly the smallest denominations, stays intensely local.

In our virtual era, smelling cash in sizable quantities has become a class-specific experience. Cash *would* smell differently if you're tallying a register full of it that doesn't belong to you. If unavoidable contact with it aggravates your skin. If, by counting it, you're also counting all the hours and wildly variable human interactions it took to earn it.

GASOLINE

Sniffing gasoline delivers a bodily hit, headlong and thrilling. The scent thrums. This smell is so big it seems to occupy three-dimensional space like a buzzing electron field.

It combines notes of honeyed naturalness with something synthetic, almost digitized. Gasoline releases its fumes in vaporous, deeply appealing waves, and—much like watching a car crash happen—one is loath to turn away from it.

Is sniffing gasoline bad for your health? In a word, yes. In a few other words: *yes, very, VERY bad.* The most appealing smell-molecule in gasoline is benzene, a known carcinogen. It's responsible for gasoline's sweet, wildflower top note.[1]

Inhaling gasoline suppresses your central nervous system—it basically makes you drunk but much, much worse. In this sense, inhaling gasoline resembles addictions like huffing solvents, nitrites, or aerosols (aka the "whippets" of my childhood, the compressed gases inside cans of whipped cream). Sniff too much gasoline, and you'll likely feel euphoric but also irritable, you'll lose coordination, your speech will slur, you might get dizzy and headachy. Sniff even more and you could cause seizures or hallucinations, lose consciousness, or even die.[2] In case it's not abundantly clear: smelling gasoline intentionally is a terrible idea.

And yet. Stealing whiffs of this scent is a vice many people revel in. It's not wrong to call this a smell fetish—gasoline mixes danger with pleasure, and it's certainly a niche preference. Yet the smells we love most tend to ride on desire as much as on air, and the truer appeal of gasoline's smell may be how well it captures a yearning to travel. This smell is redolent of car culture, of propulsive dreams of motion, of dropping twenty bucks in a tank and getting the hell out of Dodge. You don't smell gasoline in Europe or Asia in quite the same way. Gasoline is a vintage Americana smell, for good *and* for ill. It mixes personal pleasures and agency with cancer, addiction, and environmental disaster. Loving this smell has gotten more complicated—and, if we're being strictly honest, hopeless tangles are nearly always intoxicating.

It feels like an iconic smell, unchanging or changing slowly over the decades—and yet not all gasoline smells alike. This apparently singular smell has variation built-in from its pre-refined state to the final product, or—more accurately—products. Crude oil is a natural substance that stems from fossilized plants and animal matter. Crude oil's smell varies based on how that fossilized matter got compressed over millennia into oil. If the crude has low levels of sulfur compounds, that's called "sweet crude oil"; "sour crude oil" is higher in sulfurs. Sometimes, via accidents of geology, crude oil starts out with more molecules one finds in refined gasoline—naturally pre-refined, if you will. Other times crude oil is really, truly crude and needs more industrial conversion to transform it into gasoline.[3]

In the United States, gasoline for cars comes in twenty-plus different blends to meet overlapping state and federal guidelines; there are even more blends overseas. Blends are mixed to control the level of volatile organic compounds, or VOCs, released into the atmosphere. More VOCs evaporate from gas into the atmosphere in warmer weather, and more VOCs means more smog. Come summertime, Texans are smelling a very different blend of gasoline compared to the blend sniffed by Bay Area Californians with their stricter environmental regulations and cooler climates. Winter and summer blends of gas exist for every locality, too. In transitional weather, gas stations switch over seasonal blends in an uncoordinated fashion. So during an autumnal cold snap one gas station might already be selling winter-blend gasoline while the station across the street is still selling summer blend.[4]

Many other smell fetishes, like a yen for gasoline, carry a heady blend of danger, intoxication, and a sense of incipient action, with dashes of childhood nostalgia thrown in. Fresh asphalt is another petroleum product whose smell offers similar charms and health risks to that of gasoline. Sharpies, nail polish remover, cleaning supplies, old-fashioned ditto sheets in classrooms of yore (see chapter 7) are all solvent-based, delivering a hit of that inhalant high. Woodsmoke? It's beautiful but also a carcinogen. Fresh tennis balls reek, agreeably but toxically, of plastics and rubber off-gassing.

Focusing on these smells' dangers is small-minded; it misses the larger point. Smell is transient and punctuates fleeting moments. When something begins or ends or is wildly, temporarily present. Maybe people love the smell of fresh cardboard when they open a package or bleach as they prepare to dye their hair because they're not expected to notice the smells of those things at all. Yet those smells assert themselves powerfully, indirectly, because they underscore accelerating into the future: what's happening right now and what's coming next.

MUSK

It smells warm, living, and subtle. It seems to pulsate with quiet vividness. I'd expected musk to smell baroque and enormous, full of fiddly nobs and curlicues, but the one I smelled wasn't like that at all. Musk's scent is both profound and elusive, like the scent of your own skin amplified. You could say musk smells like the natural scent of humans perfected: what we smell like when we're happy, playful, and fully relaxed. The smell suggests tension, with a giddy calm on the other side.

In the botanical-musk blend I'm wearing, Embers & Musk by natural perfumer Mandy Aftel, musk hovers like a velvet screen in the background. The perfume's foreground is dominated by smokey pine tar, spicy pink pepper, guaiacol (a component of whiskey) with the lightest of floral and Japanese yuzu citrus notes. The musk note is difficult to separate out, but it's clearly the perfume's binding agent. It reminds me of a figure-skating champion's costume: all those glittering sequins and feathers, tracing complex designs on the body, are held in place by a barely visible mesh sheathing the skater's actual body. And it's this body itself that's astonishing, strong and beautiful—not the costume. Musk feels core to this fragrance's charms like that mesh is.

Original, animalian musk came from a red jelly harvested from the glands of a Himalayan musk deer. A rutting male deer might produce as much as thirty grams of musk, and while one didn't technically have to kill the deer to harvest this precious material, people generally did.[1] Deer musk has close cousin smells. These include civet, a honey-like anal secretion from wild cats native to Ethiopia and Indonesia, and castoreum, derived from the abdominal sacs of Canadian and Russian beavers. All of these scents are obnoxious in full concentration and become delectable only when diluted.

All were astonishingly long-lasting and thus used historically as perfume fixatives. And all these scents are no longer hunted in extremis from wild animals, but instead synthesized in labs.[2]

Musk-dominated perfumes were big in the premodern era. Musk and similar scents supplied a beautiful, fat bass note and staying power to the robust perfumes prevalent at the time. Musk perfumes gave way to floral scents in eighteenth-century Europe; changing social norms made animalian smells, including those suggestive of our own bodies, more disgusting.[3]

Islamic cultures in particular have extolled the virtues of musk, comparing the smell to various good things: high praise, untouched youth, royalty and wealth, true love. Like a secret or an inarguable truth, musk will always leak out. As noted in one passage of Islamic Persian literature, "Even if musk is kept concealed in the pod, its scent manifests in the world."[4]

The original musks gave humans a window into animalian smells in the wild, a thicket of signals humans can observe but aren't actually meant for us to parse. Smells among animals constitute a great deal of their knowledge and their communications with one another—almost, you might say, their cultures.

Male lemurs establish dominance among themselves by staging "stink-offs": males blast each other with mating smells until the weaker party runs out of fumes. He slinks off, castrated both literally (for a short while) as well as figuratively (in terms of status).[5] Birds sniff out each other's bodily smells before deciding to mate; what they're really doing is assessing each other's microbiomes.[6] (Humans do this, too, in their own way—see the skin section, chapter 6.)

Otters recognize their familiars via their anal scent glands and can exchange a kind of news with each other using these smells. These signals are so refined, otters hailing from different regions use different "smell-dialects"; they often can't parse anything from a strange otter's smell.[7] Lots of birds—but also mice, lampreys, and salmon—navigate extreme distances relying primarily on smells. They note all the distinct smells encountered on a long journey and then follow that pattern of smells in reverse; it's a double feat of both smelling and memory.[8] Elephants can even "count" using smells—that is, they use smell to compare different quantities of food with surprising accuracy.[9] Much of what animals know about their world depends on smells. Synthesizing those smells together with the evidence of other senses creates a coherent worldview.

The musk perfume I'm wearing now hints at the enormity of animal smells, but falsely. My musk isn't animalian at all but botanically derived—specifically from ambrette seeds. Synthetic musks abound from laboratories, too, creating not one singular musk but a whole family of related musky scents.

I sniff my wrist again. It smells rounded, pleasant, and full, as if the air is glittering with information—which, to most animals other than us, it is.

7 SHARP & PUNGENT

Waking up the senses further with bracing scents that grab you by the lapels or hurtle you backward in time.

ORANGES

LAVENDER

SKUNK

BEER

DITTO SHEETS

ORANGES

They smell sparkling, clean, literally zesty. Oranges invigorate the nose with a bright acidic tang, over which plays a light, balanced sweetness. It's affable and domestic, a lunchtime smell.

You smell oranges well before you get a chance to cut into them. Holding a fresh orange in your palm, the scent encircles the fruit like a halo of sunshine. Cleaving through the rind can release visible zephyrs of fragrance; those are the essential oils escaping from cells lining the peel. The smell of oranges peaks just at this moment of incipience. It promises juice, refreshment, sunshine in winter, a glimmer of hope for summer's return—and it delivers all of these. Smelling good fresh oranges brings with it an uncomplicated happiness.

The smell of oranges travels surprisingly far. You know you're nearing an orange grove by the dense citrus scent that greets you miles before the actual trees do. To visitors from chillier climates, smelling real orange trees as they perfume the air around you is intoxicating. This was especially true in the era predating air travel, but it's still true now. A whiff of paradise, the charm of exotic distance, still clings to oranges' fragrant peels.

Oranges can contain worlds, the central theme of my favorite book *Oranges* by John

McPhee. I cannot smell oranges deeply, or write about their smell, without my mind drifting in this beautiful little book's direction.

I learned from reading McPhee that citrus is a bit freakish. For starters, you can grow oranges, lemons, kumquats, limes, and grapefruits all on the same tree, simply by grafting buds of each fruit onto the same rootstock, or trunk. Oranges are actually green until the tree is exposed to sufficiently cold air. It's this chilliness that breaks down the green chlorophyll inside the fruit's peel cells and lets the orange carotenoids prevail.[1] Like all fruit, oranges continue to breathe on your kitchen counter, inhaling and exhaling gasses. The mingled breathing of different kinds of fruit, nestled together in a bowl, can alter how fast each fruit ripens or rots.[2] When you smell oranges, you're smelling the respiration of something still alive.

Another freakish fact about oranges: the best-smelling varietals, beloved by perfumers, are mostly unpalatable to eat. Take the Seville orange, also known as sour orange or *Citrus aurantium*. The first citrus to arrive in Italy and later spread across Europe, its bitter juice was squeezed over medieval banquets to cut the meat's grease. Whole sour orange fruits could only be eaten candied or turned into marmalade.

The *flowers* from sour orange trees can be steam-distilled to yield neroli, a super-expensive essential oil that's popular in perfumery to this day. My tiny bottle of neroli essential oil smells lilting and intensely floral, with only the faintest suggestion of fruit. It's named for Anna Maria de la Trémoille, Princess of Nerola, a supreme political operator of seventeenth-century Europe and an avowed neroli fan.[3] Neroli is such a desirable perfume ingredient, it spawned a more affordable cousin scent derived from the same plant. *Petitgrain* is traditionally extracted from the green immature fruit, twigs, and fresh leaves of sour orange plants.[4]

Another orange varietal beloved by perfumers is bergamot, whose fruit is pretty but bitter-tasting. Its name comes from the Turkish phrase *beg-armudi*, "the prince's pear," which references the fruit's pale green rind and pear shape.[5] Its peel is so dense with essential oil, the traditional method of extracting a bergamot's scent is simply to rub a sponge over the fruit's surface, then squeeze the wet sponge into a glass vial.

While we think of eau de cologne today as a generic term (it refers to a perfume with only 5 percent concentration of essential oils as differentiated from *eau de perfum* or *eau de toilette*[6]), eau de Cologne was once a very specific perfume—and bergamot was its chief ingredient. The Italian perfumer Giovanni Maria Farina formulated it in 1708 while living in Cologne, Germany, to evoke his native country. Farina described the fragrance in a letter to his brother: "the scent of a spring morning in Italy, of mountain narcissus and citrus blossom after rain." The fragrance combines lemon, neroli, lavender, and rosemary but leads with a green note of bergamot. Eau de Cologne literally bottled Italy's charms and became a blockbuster perfume in courts across eighteenth-century Europe. King Louis XV, Napoleon, and Mozart all wore eau de Cologne regularly. Goethe liked it so much, he kept a box of rags soaked in the scent by his writing table. He'd sniff the rags, pause, and then scribble on. One can almost sniff the bergamot in these lines written by Goethe, perhaps the most famous in all of German poetry and expressive of a generational yearning for Italy: "Do you know the land / where the citrus trees bloom?"

Farina wanted to keep eau de Cologne's scent consistent despite variations in its natural ingredients. This goal was unheard of in the preindustrial era, but Farina managed it. He went so far as to import fresh bergamot oranges directly to Cologne and extract the essential oils himself. Mothers in Cologne would bring their children to Farina's distillery, lifting them to the open windows so the children could inhale the health-giving zephyrs of fresh oranges.[7]

My bottle of bergamot essential oil smells remarkably different from neroli, or indeed from fresh oranges. It's herbaceous, a bit astringent, and moody. It pings a distant reminder that's difficult to place until one learns that bergamot is a dominant note in Earl Grey tea.

Chinotto are yet another super-bitter, great-smelling orange grown in Savona, Italy. Their trees are awkward and misshapen, but their blossoms release an extraordinary fragrance.[8] Italian sailors from the region brought chinotto on long voyages to ward off scurvy. Sliced in two and floating in a cask of seawater, the fruit lightly ferments and stays preserved for months. The seawater also softens their skins and tempers the bitter taste, although they still weren't particularly tasty.

Infinitely more delicious are green candied chinotto floating in maraschino liqueur. In bars across Savona, one used to find a tall glass jar in art nouveau style decorated with a traditionally dressed Chinese man holding a parasol. Inside the jar floated the chinotto, which you would fish out on a winter afternoon and use to chase your espresso.[9] Chinotto syrup also forms the basis for San Pellegrino Chinotto soda, Fascist Italy's answer to American Coca-Cola.[10] I wasn't able to lay my hands on candied chinotto fruit, but I did swig a bottle of Chinotto soda. It's dark, bittersweet, and herbaceous, like an effervescent *amaro* you'd blend into cocktails.

Oranges originated in China, but not all variants of the fruit have penetrated the West equally. Yuzu is a delicate fruit that's a hybrid of mandarin oranges and *Citrus ichangensis,* a hardy citron from southwestern China. Yuzu can be eaten fresh or cooked, and its smell is particularly celebrated in a Japanese ritual hot bath known as *yuzuyu*. Performed at the winter solstice, yuzuyu is supposed to fortify one's health against the coming winter. You drop whole yuzu fruits into hot water, where they bob gently and scent the exhaling steam. You can also scrub the nubbly yuzu rind against your skin to exfoliate, releasing even more perfume. I can imagine it's lovely just by sniffing my yuzu essential oil. The scent is rousingly fresh and clean, floral without the saccharine edge of neroli.[11]

Sniffing all these different orange smells bring me back to McPhee's book. *Oranges* examines the fruit from many angles but doesn't actually probe their smell too much. So why does it speak so much to me as I write this? What's the smell connection?

It's the writing, a whiff of what quiet epiphanies a book can bring. *Oranges* started as a *New Yorker* article about a mundane topic: the rising popularity of frozen orange juice concentrate in 1960s America. But out of this quotidian mud, McPhee reaches for the sublime and grasps it over and over. His keen interest in oranges propels the book forward. This is ostensibly a book about oranges, but it's also a book about deep curiosity. Linking facts associatively, the narrative is borne aloft like a scent twisting through a vast tract of air. How far can it possibly travel before it disperses into nothingness? You're amazed, you're stultified, you're sniffing avidly like a bloodhound. The book's many charms emerge fleetingly and thrilling, just like pinning down a smell.

Whenever I read *Oranges* again, and I've read it plenty of times, my perceptions feel supercharged, as if I'm momentarily capable of sensing the vibrations between electrons. Perhaps the most evocative moments to me in this book are the section breaks, the pauses to catch one's breath. That's where the atmosphere collects and deepens. *Oranges* embodies, for me, a concept perfected by perfumers: the miracle of writerly distillation.

LAVENDER

First comes a frilly whiff of flowers, brief and headlong. Only then comes the muscle.

Lavender's smell deceives with its potency. It's no wonder lavender symbolized mistrust to ancient Romans, although their particular worry stemmed from a belief that deadly asps lurked inside lavender bushes.[1]

The smell opens icily, almost cruelly, like a packed snowball crushed to the nose. But the crystals melt rapidly, and a faint floral note emerges. It's liquid and nectar-sweet, almost frivolous. As the smell opens up further, a very different kind of pungency barrels through. The heart of lavender's smell is spicy and herbaceous, sun-kissed with a lingering chill, and gets warmer and more enveloping as it dries down on your wrist.

Lavender's smell is strong enough to create its own atmosphere—to insist upon it, really. Splash a fair amount of lavender oil on your wrists, as I did, and you can't think about much else afterward. But crucially, you don't want to either. Its scent fills the nose with intriguing contrasts, all vigorous and imperfectly resolved: floral versus balsamic, hot versus cold. Like its cousin smells eucalyptus and camphor, lavender's scent is powerfully, nose-scrubbingly clean. Only a filigree of flowers makes lavender's scent seem prettier than the others.

Lavender's smell is comforting because it combines distraction, pleasure, and force, each to an extreme degree. By demanding that you surrender to it, lavender rinses you clean of any other anxieties. Small wonder that this scent is highly associated with the maternal: both are equal parts gentle and tough.

As a plant, lavender is hardy, ubiquitous, and cheap, yet still widely considered beautiful. Every bit of the plant both looks and smells good, and nearly everyone has both smelled it and liked that smell. In our visually saturated modern era, lavender's popularity is further helped by the fact that most people recognize the flower and can readily picture it in their minds. This is a smell we also know how to see.

The plant's Latin name *Lavandula vera* resembles the Latin word *lavere*, "to wash," suggesting a common root. Although the two words aren't related etymologically, a long tradition of using lavender to perfume bathwater or scent laundry has rendered the two words hopelessly intertwined.[2] Storing dried lavender with freshly laundered clothes helps preserve their clean scent. According to the Oxford English Dictionary, "to lay up in lavender" has several meanings, from the literal to the more metaphorical: to lay something precious aside for future use; to pawn something valuable; or to remove someone firmly from a temptation to do harm. Like a big brother–style grandmother, lavender combines prudence with a touch of brawn.

TUCKING LAVENDER SPRIGS INTO YOUR BED SHEETS makes them smell good. Does it also help you relax and sleep more soundly? Folk medicine, aromatherapy, and, increasingly, modern medicine are converging on an answer: yes. In fact, lavender is the original scent that spawned the field of aromatherapy, and its medicinal effects are well substantiated by modern scientific evidence.

Aromatherapy began as an accident, with lavender as the unlikely hero scent. Early in the twentieth century a cosmetic chemist named René-Maurice Gattefossé was mixing a new perfume and burned himself. Seeking quick relief, he plunged his arm into

EXERCISE

Alternate nostrils while sniffing.

Try sniffing a new smell first with one nostril, then the other, then both. "One of your nostrils is a low-flow nostril. It is not worse; it just works harder or longer," writes Alexandra Horowitz, author of *Being a Dog: Following the Dog Into a World of Smell.*[*] In this book, the lead researcher at the Dog Cognition Lab at Barnard College attempts to get inside a dog's head by encountering the world as they do, smells-first.

Dogs approach new smells cautiously, right nostril first; if the smell is neutral or pleasant, they'll switch to sniffing with the left nostril. (A dog that sticks strictly to right-nostril-sniffing is feeling suspicious.[†]) In this way dogs build a stereoscopic impression of a smell, much as our two eyes allow us to merge two visual perspectives into a 3D image.

If a smell you're sniffing is handheld, try rotating it toward one nostril, then the other, to approach it from a few different angles. See how many different scent notes you can observe. Imagine these scents creating a three-dimensional structure: Which scents feel like they form the smell's base? Which are the middle layers and top notes?

---

[*] Horowitz, *Being a Dog*, 118.

[†] Horowitz, *Being a Dog*, 36.

the first cooling liquid he had on hand, lavender oil. His pain abated almost immediately and, he claimed, his recovery was accelerated thanks to lavender oil. Afterward Gattefossé devoted himself to studying the medicinal effects of essential oils for the next fifty years. His book *Aromatherapie*, published in 1928, coined the term and launched aromatherapy as a cultural phenomenon.[3]

Lavender can supposedly calm anxiety, promote relaxation, and encourage sleep—all areas where many of us could use help nowadays.[4] A growing body of scientific studies—all modest in scale but numerous—suggest lavender may indeed have superpowers. In one study, lavender oil beat the placebo handily in helping participants suffering with anxiety disorder; participants slept better and felt less restless and out of sorts. Lavender helped quell anxiety and depression in a group of high-risk postpartum mothers. It allayed the fears of dental patients awaiting potentially painful treatments. It soothed the low-anxiety nerves of people who'd just watched anxiety-inducing film clips. In perhaps the most persuasive study of lavender's impact on humans, lavender oil taken orally improved generalized anxiety disorder as much as taking 0.5 g of a well-known antianxiety drug lorazepam (Ativan).[5]

Sniffing lavender myself, I found my mind straying to one animal study. In it mice inhaled linalool—a smell molecule prominent in lavender's scent—and chilled out to a degree equivalent to that of popping a Valium, but without that drug's woozy side effects. Interestingly, the mice only truly relaxed when they *inhaled* the lavender scent—injecting linalool directly into their bloodstreams did not work. They needed their olfactory receptors in the nose tickled by linalool to relax properly.

The value of aromatherapy might be incredibly simple: stopping to smell the roses—or lavender—but taking that pause seriously. After all, a smell isn't very different from a mantra in guided meditation. It's the focus for your straying attention. Smelling requires you to breathe, slow down, contemplate something usually unnoticed but lovely, while you experience the effects of doing this on your mind and body. But unlike a mantra you can repeat ad infinitum during meditation, smells are fleeting. You can only sniff and contemplate a few times before your nose gets exhausted and you can't detect the smell anymore. Meditation for me is similar: letting intruding thoughts go, wandering but returning, a practice that demands both concentration and looseness.

HERE, THEN, IS LAVENDER: clean, brusque, mind-clearing, bossy, and soothing in a maternal way, its prettiness belying its strength. Sniffing my wrists again, lavender feels sneakily feminine to me: the way Ginger Rogers performed all the dance tricks Fred Astaire did, just in high heels and backward. Lavender smells like soft power.

SKUNK

It's keen, almost knife-edged, and rank. For me the scent leads with an oily sheen of rotten-egg sulfurs, shimmering with awfulness and depth. Skunk smell combines unpleasantness, complexity, and pungency in equal measure.

It's bigger than most smells, denser and more physical. It's so concentrated it seems almost phosphorescent. Thinking about skunkiness, I picture the smell the way cartoonists draw any stink: as sickly-green vapors that thread up characters' nostrils to extreme and comic effect.

The smell of skunk persists to an incredible degree, *and* it sprawls. I can't think of many other stenches you can detect on a highway and then keep on smelling for thirty seconds while driving 75 mph away from them. Why is a skunk's smell so durable, pervasive, and intense? Do other skunks like this smell? What explains its partisan appeal to some people?

Like other nocturnal creatures, skunks operate in a light-deprived world in a very smell-forward way. Communication, tracking, navigation—they accomplish all of these largely via their sense of smell.

Skunks have only one natural defense against predators, but it's a doozy: they are precision squirters of a noxious spray. They've got two nipple squirters on either side of their anus that can rotate independently for more accurate aim, which they usually train on their attacker's eyes and can pinpoint from ten feet away. Skunk spray's effects on their victims are equivalent to those of tear gas, producing gagging and coughing, stinging eyes, temporary blindness, chemical burns, and nausea. In rare instances this stink has even killed people. Even for those who don't suffer these symptoms, skunk smell is very trigeminal, triggering nerves in your face akin to physical touch. It's really the skunk smell's unpleasant physicality that makes this stink so potent.[1]

Which isn't to say the underlying smell isn't gross, too. Skunk smell is heavy on thiols, organic compounds rich in smelly sulfurs. Thiols include the compound added to natural gas so you can notice otherwise odorless leaks of it. Thiols also appear in anesthetics and antispasmodics, partly explaining their disabling physical effects on the skunk's predators. Skunk spray is optimized with concentrated oils, so it travels far and lingers long. The smell's purpose is not just to repel an attacker, but to issue a warning detectable for miles around: the skunk who lives here is not to be messed with.

It can take up to ten days for a skunk's anal glands to refill, during which they're quite vulnerable to predators. So skunks conserve their precious stink as a last resort. If their zigzaggy stripes don't warn away a predator, they will stamp their feet noisily. If that's ineffective, threatened skunks do a crazy handstand with their tail straight up, anal glands flashing dangerously. Only if those deterrents don't work do skunks move on to the nuclear option and spray.[2]

What about mating stinks? I wondered if skunks emitted a different smell for mating, a more attractive one—much like how the chemical composition of human tears is subtly different if we're crying from physical versus emotional pain. But with skunks: no. Nobody likes the skunk smell, not even other skunks in heat. Female skunks may spray to fend off unwanted males from trying to mate with them.[3] So mating stink is a deterrent, the same smell emitted when skunks are defending themselves from attack. Skunk smell isn't romantic, but arguably it *is* feminist.

Some people actually enjoy the smell of skunk. Why? A few reasons. First, not everyone can smell the full bouquet of rich, nasty notes it contains. That's true of any scent, in fact—not all of us detect every nuance equally.

Almost nobody likes the smell of a skunk's full-frontal attack. Those who do like the scent tend to enjoy it obliquely. They'll catch a whiff floating in the winter woodland air and briefly marvel at the scent's complexity. They'll sniff a hint of it on their dog, returned from hunting, and revel in momentary wildness. The scent signals a kingdom of outdoor space, the skunk claiming her domain. It suggests all the frenzied plots unspooling between animals in nature, an ongoing show indifferent to human audiences. It's thoroughly uncivilized, jolie laide. Locating beauty within a stink is its own perverse, and very human, pleasure.

BEER

The smell is clean, sunny, complex, even vaulted. In front of me sits a witbier freshly poured into a stemless wine glass, condensation already beading on the can next to it. But my mind refuses to parse the smell further. It's too complete, too perfect to analyze.

This scent teleports me straight back to a scene I've encountered hundreds of times before: a bower of tree limbs tangled overhead with sunlight filtering through, winking with the changeable movement of leaves. A sense of civilized languor settling. The distant sounds of happy children crunching over fine gravel with its own, highly specific acoustic. Murmured snatches of spoken German that reveal the language's cozier side. The satisfaction of finding a level spot for your glass on a warped wooden picnic table. The near stoppage of time that only a late summer biergarten can deliver. For me this scene is replete with happiness.

For nearly every summer for fifteen years, we've repeated this scene somewhere in Berlin, once a week or more. This smell—of high-quality beer warming in a glass—contains within itself a *mise-en-abîme* of many unhurried drinking sessions past. How the golden daylight hours stretch ineluctably into evening. Meeting for a drink in the biergarten is a perfect social ritual: spontaneous and yet formalized enough to register mutual importance. Germans are meticulous about toasting: once everyone is supplied with a full glass and before the first sip, you clink glasses with each person at the table, no matter how large the group. You always make eye contact; you never cross arms with other toasters. Then you settle back and spend time together, at least the interval it takes to drink a beer or two. This scene unfolds from the vapors escaping my glass now like magical origami.

The pandemic summer of 2020 is very different. I'm sniffing this beer at 11:30 a.m. from inside an office tower on Chicago's Magnificent Mile. To get here I picked my way past boarded-up storefronts, around crushed glass on the sidewalk. I can remove my surgical mask inside the locked office to sip and sniff. Here and now is one reality, the beer powerfully suggestive of another.

The spell it casts is temporary; it's almost a relief to dispel it by analyzing the beer's smells. If this beer teleports me forcefully back in both time and location, its smells can also help me time-travel back into the brewing process. An expert can reverse-engineer a lot of how a beer was brewed from its smells and other sensory cues. I've enlisted such an expert via email: Mikey Manning, head brewer at Marz Community Brewing. I'm drinking his handiwork, an Orange Velour Belgian witbier.

Beer is made of four ingredients—water, yeast, hops, and malt—in four classical stages. The first stage is mashing: you take partially sprouted, or malted, grain like barley and douse it with hot water. This releases the malt's sugars, creating a slurry called wort. Witbiers start with wort made of malted wheat and barley. It's this mash that gives beer its specific color and foundational qualities and smells. My witbier smells clean, crisp, and sweet, in tune with its signature honeyed color. A more roasted, caramel, or chocolate scent would've revealed this to be a dark beer even if I were blindfolded.

Step two is boiling, which pasteurizes and concentrates the wort. This is also the stage at which brewers typically add hops, strategically altering the beer's bitterness, flavor, and aromas. Adding hops involves a simple question of timing and a much more complex one of which hops to choose. First, on timing: If you're adding hops with an eye toward imparting more bitterness to the beer, you add it early in the boiling phase. If your aim is to infuse more flavors and aromas, then you add them later so you don't boil off the hops' volatile oils. Hops offer a galaxy of different smells: fruity notes like peaches, lemon, oranges, melon, mango, and other tropical fruits; floral characteristics like fresh-cut grass, flowers, lemongrass, or other herbs. My witbier is redolent of zested citrus peel and coriander, I'm guessing added later in the boil.[1]

Step three is fermentation, which starts by cooling the boiled wort and then adding (or "pitching") the yeast into it. The yeasts convert the wort's sugars into alcohols and produce carbon dioxide as a byproduct. In other words yeast unlocks beer's more overtly fun qualities, booze and effervescence. An anaerobic fungus, yeast is earthy-smelling and releases different aromas during the fermentation stage. Sipping the final product, I can't smell the yeast directly the way someone with Mikey's training might, but yeast's aftereffects are clear: a frothy pour with appealingly tight bubbles, releasing top notes of banana and cloves.

Beer-making Vikings, unaware of the chemical importance of yeasts to beer, used to swear by the family's brewing stick to stir the wort. Turns out the spoon teemed with that family's distinctive yeasts.[2] In this sense beer-making resembles cheese-making in that friendly microbes quietly invade a cave, or a wooden spoon, and do their magic unawares.

Last, in step four, the beer is stored. Sometimes it's poured into bottles with an added flip of sugar or extra yeast for a second fermentation. Other times it's aged in wooden barrels, then bottled for consumption. My witbier arrived in an aluminum can; all its smells derived from earlier stages. But a barrel-aged beer might absorb the characteristics of the barrel, whether oak or sherry cask or whiskey cask. A fruit sour might alter in color, acidity, and scent with the addition of fruits in the storage phase.[3]

Stinks crop up not infrequently in brewing due to oxidation. Yeasts can unleash many flavor and aroma compounds in beer, one of which is the ketone compound diacetyl. It gives Scotch ales, stouts, and heavier pale ales a desirable buttery note, but diacetyl is an unstable molecule and can decompose rapidly into a stale or raunchy smell offensive in light lagers. In a process called kettle souring, the brewer inoculates the wort with a lactobacillus culture, lowering the beer's pH level and making it sour or tart. If the wort is exposed to oxygen during this delicate process, a butyric stink will invade the beer—redolent of baby vomit or blue cheese.

The threat of encroaching stinks makes brewers fanatical sanitizers, particularly for "clean" styles like lagers and pale ales. They also choose their yeasts carefully for the compounds each tends to produce. Every once in a while, though, brewers will channel wild yeasts riding invisibly in the air. Beers that call for environmental yeasts are brewed in an open-mouthed vessel called a *Koelschip*, traditionally during the winter months in an unheated barn or a room with the windows flung open. Natural yeasts float into the vessel from the air, which brewers sometimes supplement with additional yeasts or house bacteria cultures.

For me, a beer's true character emerges at room temperature. Good beer rarely lasts long, but biergarten time is sweetly distracted. Time will pass, and beers will warm in the sun. You can protect your glass from intruding wasps—August is *Wespenzeit*, wasp season, in Berlin—by topping it with a cardboard coaster when you're not actively drinking.

Back in pandemic Chicago, I got distracted from my own beer contemplation by several midday conference calls. When I returned to my Orange Velour an hour later, it betrayed no detectable fizz when I swirled the glass. But one sip revealed an explosion of microbubbles on the tongue, a surprisingly belated fireworks. Time should always be so forgiving as it passes.

DITTO SHEETS

It's a smell unrecoverable now, except via memory. And so the mind reaches backward to childhood and school days.

The student seated nearest to the classroom door would receive a stack of paper, warm and lambent in the fingers. You'd peel off your own sheet, pass the stack down the aisle, then wrap the page around your cheeks and greedily sniff. Ditto-sheet smell would infiltrate the classroom with this sudden aeration of pages, like a huge fragrant bird flapping. And it vanished almost instantly.

To an uncanny degree the smell merged with how a dittoed page looked, maybe because that's all you were left with after the scent evaporated. A faint wash of purple type, slightly blurred, wandering off-center inside a crisp white rectangle of paper. It preserved a record of its own production. You could discern the ditto sheet's recent, wet past in its present dry self.

Dittos were copies, and distinctly mechanical ones. (The word *ditto* comes from the Latin *dictus*, "having been said," the past participle of the verb *dīcere*.[1]) One could see where the typewriter's keys bit more heavily into the master sheet's paper, or where the writer pressed down unevenly on a pencil if the master sheet happened to be handwritten. Some copies in the stack would be fully saturated in deep purple ink; their letters seemed to swim to the paper's surface, startlingly high-res and legible. Other copies would be paler and ghostly like passing thoughts: these were the final sheets in the run.

For me ditto sheets play an interesting havoc with the idea of aura, a quality the philosopher Walter Benjamin claimed only original artworks could have and copies could not. He defines aura as "a strange tissue of space and time: the unique apparition of a distance, however near it may be."[2]

Ditto sheets never claimed to be copies of "artworks" at all. You'd ditto study sheets, permission slips for class trips, school newsletters, pop quizzes. And yet ditto sheets— cheap, unartistic copies run up in schools and church basements—managed to embody aura in their very imperfection. Even in childhood, a ditto sheet elevated whatever mundane message it contained. A scrim or veil seemed to hang between your copy and the master sheet. You'd write your name at the top of the paper, then start filling in the blanks provided. Your handwritten scrawl would float above the spectral purple typesetting: yet another veil of distance. The haze of memory now produces a third layer in this veil.

Benjamin gave an example of the aura from nature: "To follow with the eye—while resting on a summer afternoon—a mountain range on the horizon or a branch that casts its shadow on the beholder is to breathe the aura of those mountains, of that branch."[3] The ditto's smell—heady, evanescent—seemed to float inside this distance.

It both pleases and spooks me to learn that ditto machines produced copies via a technology known as the "spirit duplicator." For ditto machines, these spirits were methanol and isopropanol, two alcohol-based solvents responsible for the scent. To create the master copy, you'd write, draw, or type on a two-ply sheet. One sheet was just ordinary paper and the other was coated with wax impregnated with a colorant, usually aniline mauve. The pressure of writing on the first sheet transferred the colored wax from the second sheet to the coated backside of the first sheet, producing a raised mirror image. You'd then hand-crank paper through the printer, where an absorbent wick spread a pungent-smelling solvent across each sheet. The blank paper would come in contact with the waxed original, removing just enough pigment to copy its image onto the sheet. You could run fifty copies before the waxed original's pigment would get used up.[4]

Remembering ditto sheets' smell gave me a jolting realization: the smells of classrooms have quietly transformed over the last thirty years. Allergy-inducing chalk dust has mostly given way to the tang of dry-erase markers. Transparency machines and 16 mm movie projectors no longer ping their smell of brittle, melting plastic and baked-on dust in the cool darkness of a classroom. Among these olfactory losses, the perfume of warm purple ditto sheets seems the cruelest. These smells join other bygone classroom vapors, even older. This may be apocryphal but I read somewhere that, as late as the 1960s, some kids repurposed old cigar boxes to carry their school supplies. Imagine it: to begin the day's work, students would flip the box lids and release a waft of aged tobacco.

Ditto sheets were big during my elementary years in the 1980s, but they trail off in memory during high school, and I don't recall encountering them at all in college. As it turned out, my elementary school days coincided with the end of spirit duplicators' decades-long run. Invented in 1923, ditto machines were technically supplanted by digital duplicators like the Xerox in the mid-1980s. But dittos were solidly built and stuck around for as long as solvent supplies lasted—the latter were eventually replaced with odorless, nontoxic variants.[5] Technologies stretch and overlap, casting their old-fashioned auras far into the future. But never indefinitely so.

As I've aged, I see now how days come and go like ditto copies of each other, blurry and interesting and mostly similar, until a particular phase of life—grade school, college, the years of living on Mulberry Street in New York, your kid's toddler years—gets suddenly used up. That particular run has ended, the ink depleted. Time for a fresh master. Like ditto sheets, the perfume of those lost days is most evident at the tail end; it's most insistently clear just as it evaporates.

8 SALTY & NUTTY

A moody and atmospheric chapter, contemplating vast oceanic smells as well as smells that primally suggest home.

OCEAN

AMBERGRIS

PLAY-DOH

WET WOOL

PEANUT BUTTER

It's briny, equal parts fresh and rotten, and rolling. The scent is decipherable into recognizable components—seaweed, scum, clams—but also insistently singular. Ocean air always seems to move, enveloping you. Its smell soothes you, lulls you to sleep, makes you peckish, exerts a forceful change of mindset. This smell is so big, you can't call it a smell anymore once you're inside its ambit. Air transformed by the ocean makes its own enormous atmosphere.

When you first approach the ocean, all your senses know it—but arguably it's your sense of smell leading that impression and crowning it. Even from fifty miles away, the ocean's smell asserts its presence. Smell is the ocean's leading signifier, its most visceral and convincing one.

The smell of the ocean is monolithic yet fluctuating, revealing activities hidden in the ocean's depths. Dimethyl sulfide is the ocean's funkiest note: sulfurous, green, clammy, touched with salt and sweet creaminess. This stink heightens when phytoplankton die in droves; the odor attracts bacteria, fish, and birds alike to feast on them. A strong concentration of dictyopterenes—the signature smell-note of seaweed—prevails whenever seaweed attempt to reproduce. Floating in the vast ocean, seaweed eggs emit this smell to help seaweed sperm find and fertilize them. Smells *can* travel in water just as they do in air—smells guide spawning salmon to their birthplaces, for instance. Smells move with water currents, disperse and weaken in wide-open oceans, and collect and intensify around obstacles like coral reefs. Thinking about smells in water seems strange, but it can help you more accurately picture their movement in air.

The brininess characterizing wild-caught fish arises from bromophenols, another ocean smell. Emitted by marine worms, algae, and bottom-feeders that fish eat, these compounds subtly flavor the fish. We eat, and smell, what they have eaten and smelled in a kind of ghostly palimpsest.[1] It reminds you that millions of years ago, humans' ancestors emerged from the sea. No wonder the ocean is both estranging and comforting: it's pinging an extremely old welcome of home.

Perfumers have bottled the ocean's smell with varying degrees of success. I bought some onycha tincture, a natural perfume ingredient made from mollusk shells. Despite much sniffing around I couldn't catch any ocean whiff from it, only the alcohol it's suspended in. A rust-red sediment swirls and clouds the liquid in my tincture bottle. Onycha combined with galbanum, stacte (see the frankincense and myrrh sections, chapter 5) probably comprised an incense called ketoret that perfumed King Solomon's temple. Onchya is supposed to smell leathery and animalic, a good grounding base note for ethereal perfumes.[2] In the modern era, chemists attempting to synthesize a new antianxiety drug failed at that goal—but the compound they did make smelled great: oceanic, melon-y, with notes of metal, anise, and a fattiness like milky oysters. Later trademarked as Calone, this synthetic molecule now imparts an exhilarating marine note to perfumes, shower gels, and detergents.[3] (For more on invented smells, see chapter 10.)

You can fathom the ocean only shallowly, but its smell brings some of that vast mystery ashore. In its scent we can briefly inhabit the ocean's depths, aerated above the water's surface for our puny mammalian lungs. The ocean's smell resists parsing. Rather it hints at a larger mystery: all the teeming marine life we cannot directly know.

AMBERGRIS

Four pebbles, in various shades of gray, sit in front of me. The first three hail from New Zealand and arrived in my mailbox in a tiny drawstring purse. Each is wrapped in tissue paper and labeled 1, 2, and 3. The fourth pebble arrived by way of a Canadian trader in a manila envelope; the lump itself was wrapped in tinfoil.

They smell worlds apart. The New Zealand trio release a burst of brininess as they tumble from their purse. At first you smell only salty top notes, giving way to crab guts, rotting seaweed, wet sand, and tar. It's beachiness on a moody off-day, roiled by storms and inclement weather. If you blink and sniff again, you can teleport yourself into a barnyard. Crushed hay, sweat-lathered bodies of animals, manure. My trio of lumps are numbered to indicate their respective grades. Number 3 is dark slate-gray and the most gamey and rich in smell. Number 1 is a pale chalky blond with a scent poised between tidal salt, fresh-shucked oysters, and a vase of overblown flowers. Number 2 sits neatly between these two.

It's the fourth lump, the Canadian one wrapped in tinfoil, that proves revelatory. Its dark mottled exterior is inauspicious; it doesn't look like fancy-grade ambergris to me. But one sniff fills the nostrils with a transcendent, complex chord you'd never expect from a rock. Ambergris's smell is often likened to quality tobacco, the wood in old churches, damp woodsy ferns, a dash of violets. All true statements, and yet incomplete. Following the instructions supplied by one of my ambergris traders, I decide to apply the "hot needle test" to my Canadian lump. This involves applying a lit match to a pinhead until it glows red, then plunging the pin into the pebble. The ambergris melts on contact with an audible hiss and plume of smoke, and it briefly exhales a stronger jet of scent. In that instant when it fills the room, the air feels disturbed. It's dizzying, amber and flower-filled, like spinning oneself in circles while gazing up at a masterful ceiling fresco in a church. Who cares about the crush of sticky-fingered tourists surrounding you, the sickly sweet altar flowers, and the dripping candles? It's an impossibly beautiful intrusion of another realm into this one.

Ambergris is a perfume ingredient of extraordinary costliness, regularly outstripping the price of gold. Its formation in nature is a near-total mystery and a literal shit-into-gold tale.

Chapter 92 of *Moby-Dick* describes the origin of ambergris as "a problem to the learned," speculating that it had something to do with "dyspepsia in the whale."[1] That's more or less accurate, but we haven't learned a lot more about ambergris's origins since the 1850s. Here's what we do know. An estimated 350,000 sperm whales are alive today and they swim in deeper waters of all the oceans on earth, making them difficult for scientists to study up close. As they roam, the whales eat a ton of squids and cuttlefish every day. The beaks of these animals can get stuck in the whale's digestive tract, and poop[2] accretes around the beaks like kidney stones.

In a process nobody understands because no human has ever witnessed it, these blockages can burst forth from the whales—whether painfully or not, killing the whale in the process or not, is unknown—and then float into the open sea. The stones are cured by the ocean, slowly turning into ambergris and eventually washing ashore.[3]

Ambergris's names in various languages reflect the mystery around its origins. Known to Muslim and Persian traders since 700 CE, European sailors first encountered ambergris in the fifteenth century as the Spanish and Portuguese trawled their way from the coasts of Africa eastward into the Pacific.[4]

Europeans adopted the French name ambergris, or "gray amber," for the substance's likeness to fossilized Baltic tree amber. By the sixteenth century, the Chinese had termed it *lung sien hang*, or "dragon's spittle fragrance." They believed ambergris was dragon spit that dropped into the sea and then solidified. Persians called manatees *gawi-ambar*, "ambergris cow," stemming from an incorrect belief that the substance came from manatees. Only the Japanese name, *kujira no fu* or "whale dung," hints at ambergris' correct source.[5]

EXERCISE

Vary your sniffing.

You may catch inadvertent smells while simply breathing. But breathing alone does not an intentional smeller make. For that you'll need to learn to sniff.

Ernst Heinrich Weber, a pioneer of modern experimental psychology, wondered how important sniffing is to detecting odors. To test his ideas, Weber laid down on the floor and had his lab assistant pour a solution of water and cologne directly into his nostrils. Sputtering delightedly to his notebook, he concluded that passively introducing a strong smell into the nose was not enough for detection; you need to *sniff* to detect any smell. Later scientists further tested the value of sniffing by introducing smells in various odd situations, all of which precluded sniffing. One scientist wafted a pungent odor in front of someone expelling air hard through their mouth. Another trickster injected odors intravenously into people with sleep apnea, between breaths.[*] All these experiments yielded the same basic result: if you're not sniffing, it's difficult to impossible to smell anything.

So sniff more, and vary your approach to sniffing. Bigger sniffs don't always give you a fuller smell-impression.

[*] Horowitz, *Being a Dog*, 38

Ambergris has been prized for centuries to a degree that's difficult to summarize (or exaggerate). It signals luxury and aphrodisiac in its own, highly efficient shorthand. In the Koran, the virginal houris populating Paradise had bodies composed of musk, camphor, saffron, and ambergris.[6] Marie Antoinette used to host lavish dinners in which guests would ritually behead a doll representing a political adversary. When beheaded, the doll would gush forth liqueur flavored with ambergris, and ladies around the table rushed to dip their handkerchiefs in it.[7] Since 1626 all British monarchs have been anointed with ambergris at their coronation.

Ambergris starts out colored black, with a fecal smell of strong brine. Finer quality pieces look like a pumice stone and smell like seashells, overlaid with a fine sheen of flowers. One ambergris trader I spoke with said ambergris might sometimes emerge from the whale very shit-like in nature; other times it may leave the whale's body already semi-cured and higher quality. Nobody can be sure.

Certain conditions seem to help its transformation along. It may help when ambergris travels through ocean waters that aren't too cold. Always floating near the surface, ambergris should get frequent exposure to sunlight and agitation from churning waves. It's an unhurried and open-ended process: it takes however long it takes. Lumps of ambergris can bob along for weeks or even years.

Once it reaches land, ambergris is fiendishly difficult to recognize. Ambergris resembles many trashy things that wash up on beaches: tallow, sheep carcasses, dog feces, rotting seagull bones, industrial waste, old whale blubber, plain stones. But real ambergris loses much of its briny smell after a few days of drying. It might reveal visible squid beaks, or striated layers. It should float in water and feel tacky to the touch like fossilized chewing gum.[8]

Ambergris cannot be farmed or cultivated. It used to be hunted, but killing sperm whales to the brink of extinction has rightly halted this practice. You cannot toss an immature lump of ambergris into a tub of salt water garnished with seaweed and hope the sunshine and breezes will alter it any further. It can age and mellow on a shelf, but the beach is where the transformational alchemy stops and longevity begins. Ambergris's smell lasts a very long time, making it traditionally useful to perfumers as a fixative. (Modern perfumes tend to use synthesized ambergris instead.) A piece of natural ambergris can retain its scent for three hundred years.[9]

Ambergris mania is still real among beachcombers and dreamers. Its fanbase's conversations are crammed with headlines about the size of their discoveries, how much each lump was worth, promising hunting grounds, lucky superstitions to improve your odds of a big and fantastical find.

Ambergris's charm is simple but as large as oceans. It's spat up by the oceans almost anywhere in the world, tracing a dotted line from an unknown origin. It might cure for days, months, years, or decades. Its beauty is heightened by its muteness. We just don't understand it, and perhaps we never will.

PLAY-DOH

A rounded smell, satisfying yet enticing. The first scent is sun-warmed wheat, edged with damp saltiness. It's that saltiness that vaults the senses into nostalgia. I close my eyes and picture a nighttime arrival at the beach. With other senses dark and quiescent, the ocean's smell enters the nose keenly. Play-Doh's salty smell also reminds you of its doughiness, plangent and inviting to the fingers. It suggests the first stage of baking, with so much coziness and sweet invention ahead of you.

These main notes are encircled with lighter top notes. An uncomplicated vanilla, a musk. A generic floral note, just sweet enough to seem edible like rose water or cherry lollipops. Overall, Play-Doh's smell is cheerfully industrial. I'm sniffing the friendly transformation of nature into commerce—a smell so iconic it's protected by trademark.

Play-Doh was born—or, more accurately, reborn—in 1956. At the time, its manufacturer Kutol Products was a behemoth of an antiquated industry: they sold putty to wipe wallpaper clean of soot buildup. But as homes switched from dirty coal heating to cleaner gas and electrical systems, there was less and less soot to clean away. Kutol's CEO's sister-in-law thought she could use the product in her nursery school as modeling clay instead. Cleaning product shifted into hit toy with only a few tweaks: Kutol opened a new division called Rainbow Crafts Company and renamed the putty Play-Doh. They made Play-Doh available in many bright colors beyond its original white, and they added the signature smell.

I say "added" advisedly. Owning a smell for commercial purposes is a surprisingly nuanced legal business, as the story of Play-Doh demonstrates.

Kutol applied for a patent on Play-Doh in 1960 and secured it in 1965, the same year General Mills bought Rainbow Crafts. Hasbro, Play-Doh's current manufacturer, acquired the brand and its patent in 1991.[1] In May 2018, Hasbro successfully trademarked Play-Doh's smell, inciting mouth-foaming debate among Redditors and IP lawyers everywhere.

Here is where smell and law entangle in a strange fashion. *Patenting* an invention is all about a product's core functionality. If you create a product for a specific purpose, consumers will pay you for your invention's functionality—hence the need to patent it and protect that revenue. A trademark, however, relates to aspects of a product that go beyond functionality, its intangible brand attributes that together signal quality. Trademarks encompass sensory cues—visual designs, distinctive sounds, colors, and smells—that help consumers know they're buying the correct product from a trusted source, not a confusingly similar competitor.

You can patent a smell if it's core to your product's function—think air fresheners, deodorant, or breath mints. Kutol successfully patented Play-Doh as a new "composition of matter" with newly imagined functionality.[2] But it's surpassingly difficult to trademark a smell. Chanel No. 5's crystal bottle, its packaging, its brand name, its mile-high billboards, and reams of scent strips in chunky fashion magazines—all of these brand elements are trademarked and zealously defended as such. They evoke an iconic brand, with the "substantial evidence" necessary to prove that consumers seeking the real Chanel No. 5 look for these signals.

The actual perfume isn't patented but theoretically could be. Perfume does have a function: helping you smell good in a particular way. Chanel could patent No. 5's composition—but only if they're willing to disclose the exact ingredients and formula for an iconic fragrance in a patent application. The perfume industry is both ultra-secretive and illogical. They're reluctant to confirm Chanel No. 5's precise formula and current ingredients, even though mass spectrometer technology makes it fairly simple to analyze a perfume and reverse-engineer its molecular composition. The upshot is that Chanel No. 5 is not patented, and the scent itself is also not protected by trademark. Perfume is an IP legal loophole.[3]

Smell trademarks are few and far between. Only thirteen active trademarks exist, and the list is rich in gimmicks. In addition to Play-Doh, the only other national brand with an active smell trademark is Verizon, which has trademarked a "flowery musk" scent that greets visitors to their destination retail stores in major markets. Other smell trademarks include fruit-scented motor lubricants; jelly sandals that smell of bubble gum; flip-flop stores redolent of coconut; ukuleles impregnated with a piña colada scent; a strawberry-smelling toothbrush.[4] The list of rejected smell trademarks, both in America and abroad, is similarly curious: citrus-scented hydraulic fracking fluid. Darts feathers that smell of strong, bitter beer. Tennis balls perfumed with the smell of fresh-cut grass.

Play-Doh is singular in its field. It's a fully protected smell trademark, perhaps the most successful in its category. Its trademark filing describes the smell as a "combination of a sweet, slightly musty, vanilla-like fragrance, with slight overtones of cherry, and the natural smell of a salted, wheat-based dough."[5] Play-Doh smells exactly like its ingredients; it's a salted wheat dough with a pleasing dash of nontoxic fragrance. Yet this simple scent amplifies the brand in a magically additive way. The secret isn't in the smell itself. It's in you.

WET WOOL

It's musty and dank, a scent that collapses distances. Somehow the smell of wool always registers as close, sometimes uncomfortably so, even if you're smelling it from the other end of a room. It seems to draw the walls closer together.

It's a homely smell, marking the border between outdoors and indoors, between the wild wet and cozy hearth. Between feeling lonely and feeling welcomed.

What is wool, and why does it smell as it does? As a category, "wool" stretches to include many furs shorn from animals: goats, hairy cattle, camels, musk oxen, bison, and of course, sheep. Wool is distinct from ordinary fur in that the hairs are kinky—a quality known as "crimp"—and elastic, meaning they spring back into shape when stretched. Wool's tight curls trap air, which makes it a great insulator for heat. Its crimp makes it easy to spin wool fibers into yarn or, with the right kind of agitation, to "lock" them with each other, producing felt.

Wool's relationship with water is downright curious. It can absorb huge quantities of water—up to one third of its weight—and the water enters not the wool fibers themselves, but the roomy air pockets between them.[1] In a literal sense wool creates its own interior world—and inside this wooly world a lot of water can slosh. Wool's interior is deep enough to absorb sound, too.

Unsurprisingly, wool smells like sheep and the environments where they live—and that smell lingers in the final fabric despite industrial-level washing. Once the wool is sheared from the sheep, it's scoured, removing dirt, grass, bits of dung, sweat salts, and most but never all of its lanolin. Secreted by the animal's sebaceous glands, this water-proof "grease" helps the sheep shed water from its coat. (Turns out sheep are growing their own little overcoats directly out of their backs; we just borrow them and reconfigure them to our size.) Sheep farmers extract as much lanolin as they can from the wool and repurpose it for beauty products and other uses, like softening up new baseball mitts or lubricating airplane wheels. But always some lanolin lingers in finished wool. Lanolin

Sweet stink of the hookah, couscous, kebab,
exhaust fumes of a bus deadlock.

—Zadie Smith, author of *NW*

smells faintly medicinal, like petroleum jelly without the artificial, plasticky tang. It's softer, more soulful, deeply wholesome.

When you smell wet wool, really what you're smelling is a four-dimensional picture encompassing both space and time. First comes the outdoorsy tramp in wet weather. Your wool outer layer fills up, slowly but inexorably, with water. It may be pleasant, until suddenly it's not. Finally you return indoors, peel off the wool clothes and stretch them out near a heat source—say a fire or radiator—to dry out. That's when the wool smell announces itself most clearly, and conditions are honestly perfect. Coaxed forth by heat, helped along by its damp greasiness, the smell is detailed because it's releasing an entire landscape of smells, a temporal sliver of time. You're smelling wool and lanolin but also any other ambient smells that got trapped inside the wool's air pockets since it was last washed. It recapitulates journeys: to the library, the pub, the fishmonger's, the gym. Wool releases all kinds of smells at once, like a sonorous, richly complicated organ chord.

To some people, wet wool smells like England—and indeed the smell is intricately bound up in the nation's historical identity. In a book called *The Last Wolf: The Hidden Springs of Englishness,* author Robert Winder explores the forces that shaped England starting with the Gulf Stream, bringing with it the warm and abundant English rain. The rain feeds the wheat, which is transformed into the iconic English drink, ale. Rain and wheat feed England's many sheep, who turn both into wool. And that wet wool becomes a crucible for England becoming itself. Chasing out the wolves (and, less excusably, Jewish financiers) paved the way for the English policy of enclosure. This cleared vast tracts of land for sheep farming, which England converted into a brisk global wool trade, the economic basis for Britain's later empire. In a book review for the *Spectator*, Dominic Green quips: "Add coal, and Winder can reduce Englishness to a 'playful equation': $E = cw^4$ (Englishness = coal x wool, wheat and wet weather)."[2] Evoked in this way, the smell of wet wool takes on a different dankness: the smell of ancestry, perhaps the lingering stink of empire.

The smell of wet wool permeates many public spaces. Wool often upholsters the seats and carpets inside trains, buses, and aircraft for a good reason. Wool ignites at a higher temperature than cotton and synthetic fabrics; flames spread only reluctantly over its surface. It's stain-resistant and durable. Yet wet wool smells very differently at a way station, or a dreary public institution like an orphanage or a convent, than it does at home. In a public conveyance, you're temporarily dry and housed, but not truly comfortable. It only reminds you forcibly how removed you still are from divesting your sodden layers in your own private vestibule, waiting for the tea kettle to whistle. The smell measures the degrees of distance to home.

PEANUT BUTTER

It's rich, smoky, surprisingly deep. The scent stacks in clear layers: at the top floats a note of honeyed sweetness. A heavy swirl of oil forms the grounding base note. In the fat middle, it's all sticky, particulate peanuts: a smell that matches the taste with uncanny fidelity.

It's a scent clouded by childhood nostalgia, mostly for Americans middle-aged or older. Younger Americans smell peanut butter much less frequently because many people in recent generations—for reasons scientists don't entirely fathom—are allergic to peanuts. My own son has a peanut allergy, which led us to purchase WowButter, a soy nut butter substitute that he eventually palled on. Our explanation: WowButter is the uncanny valley of peanut butter. Because it's roasted in the same way as peanut butter, its taste is a near-perfect match—but that nearness is also what mars the effect. But honestly, he has zero nostalgia for peanut butter; this only explains our recoil.

Non-Americans can find the smell of peanut butter polarizing, complicated by American hegemony. Either they're repulsed, having not imprinted on peanut butter in their youth, or they may fetishize it as an acquired, imported taste. As America's influence declines abroad, peanut butter's cachet may suffer, too—but for now, this quotidian smell stretches surprisingly far, from the pantry to the outer reaches of the universe.

For starters, intense craving for the smell and taste of peanut butter can reveal deficiencies in your diet. People who eat either very low-fat or low-carb diets often crave peanut butter for good reason. Peanut butter efficiently combines high-quality fats as well as a sweet hit of nutrition-rich carbs. It also contains a specific compound, beta-sitosterol, which combats the bodily ravages of stress. Studies suggest beta-sitosterol can normalize cortisol levels and reduce feelings of anxiety.[1]

Peanut butter's smell may also reveal the onset of Alzheimer's. In one study, four groups of older people were asked to smell a spoonful of peanut butter. The first group included probable Alzheimer's patients. The second group had other mild forms of cognitive impairment, the third group had other cognitive issues, and the control group was cognitively normal. Only the probable-Alzheimer's group had trouble smelling the peanut butter. And their smelling problems were weirdly specific in a way that's telltale for this disease. The left nostril had particular difficulty smelling the peanut butter; participants either had to get really close or couldn't smell it at all with their left nostrils. The right nostril could detect a whiff of peanut butter from much farther away. Imbalanced smelling of this kind may be a hallmark of Alzheimer's that helps confirm diagnoses.[2]

Peanut butter's smell could also help fight the COVID-19 pandemic. A sudden loss of taste and smell (anosmia) is an excellent predictor of early onset or asymptomatic COVID-19, much better than temperature checks.[3] At the time of this writing, when diagnostic testing is scarce, scientists and manufacturers are jockeying to find smell tests that could be easily administered at home, work, or school. A Yale scientist particularly likes peanut butter as a test smell for North Americans. Not only is it available in most households, peanut butter is what's known as a "pure odorant"—it doesn't trigger the trigeminal nerve at all. People may be slow to recognize a decline in smelling capability, because a trigeminal nerve reaction obscures the loss. You feel the expected nasal reaction and backfill the actual smell. In the Yale scientist's proposed test, you could sniff peanut butter to test your sense of smell, then sniff vinegar or rubbing alcohol to irritate the nasal passages. If you can pick up the alcohol's sting but can't smell the peanut butter clearly, head to the doctor.[4] In the workplace and school version, people might be asked to identify four scratch-and-sniff smells in a multiple-choice format before entering the building.[5]

The morning wind spreads its fresh smell. We must get up and take that in, that wind that lets us live. Breathe before it's gone.

—Rumi

All this diagnostic sniffing of peanut butter made me wonder: can smells actually hurt you by triggering an allergy? The short answer seems to be no for most people, maybe for severely allergic people where the smell can indicate free-floating peanut dust in the air—think of those stations in specialty grocers where you can get peanut butter freshly ground from nuts. It's the peanut proteins that trigger allergies, which are less prone to becoming airborne.[6]

Now for the part where peanut butter's smell gets truly weird.

The Murchison meteorite fell to earth in Australia in 1969. To check how old this hunk of space matter is, scientists checked how many different elements the meteor contains. As it bumbles its way through the universe, space matter interacts with high-energy cosmic rays. The longer it bumbles, the more cosmic rays it encounters and the more elements form inside it. Really old space matter will reveal lots of different elements via a neon isotope test. Researchers at Chicago's Field Museum and the University of Chicago suspected the Murchison meteorite was extremely old and might even contain "pre-solar grains"—stardust that predates the formation of our own sun 4.6 billion years ago.

Before researchers could test this theory, they needed to pulverize bits of the meteorite to a powder. This yielded a paste, smelling strongly of rancid peanut butter, which they dissolved in acid to yield pure stardust. This particular stardust turned out to be even older than anticipated. At 7.5 billion years old, it's by far the oldest material ever to land on earth.[7]

In my first book, *ROY G. BIV*, I wrote about a group of astronomers who, as a lark, attempted to calculate the average color of the universe. They determined the universe is colored beige—a shade they dubbed "Cosmic Latte." And now we can answer semi-scientifically yet another odd question: what does the oldest material in space smell like? And that answer is: like rancid peanut butter.

TINGLING & FRESH

Time for an invigorating stroll. All these smells are mind-clearing and bright. Understanding their histories and effects on mind and body is similarly illuminating.

SNOW

GINGER

ROSEMARY

PINE

SNOW

Under a heavy white sky, you inhale the chilled air and pause. Between inhale and exhale, inside this pause, the smell emerges as clear as thought. It's colorless but insistent. It lacks all other qualities but conviction: very soon, it will snow. But what makes your nose so certain?

This smell registers like bodily knowledge because, in a sense, it is. Your nose senses oncoming snow in three ways. When all three come together, they form a recognizable signature. First of all you're smelling cold air, dense and logy. Coldness slows the air's molecular activity. When it's cold outside, volatile molecules stay put, refuse to become airborne as scents. There's noticeably less in cold air you *can* smell. Your nose is registering blankness, an unusually complete absence of smells.

Amid this blankness, your nose observes a second something: rising humidity. Like all precipitation, snow results when the air fills with moisture until it reaches maximum saturation and falls as snow. Your smeller works best in warm, humid conditions. In balmy weather, rising humidity is often obscured by showier smells. But during a cold snap, a sudden humidity bump switches your nose's smelling capabilities back on. It registers vividly, like a jet of technicolor in a deadened landscape.

The third way you smell snow isn't smelling at all, but a related physical sensation.

Cold, humid air stimulates the trigeminal nerves in your face and nose. You're not so much smelling the oncoming snow as feeling it inside your face.[1]

Once snow has actually fallen, its smell shifts back to blankness and brings a contemplative emptiness. Take a walk over a fresh blanket of snow and sniff. With your nostrils cleared of competing scents, any smell that emerges does so keenly: your breath condensing wetly on a wool scarf, a sudden shower of pine needles releasing a waft of resin, a fine thread of woodsmoke drifting from miles away.

The smell of oncoming snow is real. So, too, is the aural hush that accompanies a fresh snowfall. Air trapped between individual snow crystals absorbs sound and changes the acoustic. For a brief time, a snowy landscape feels spare yet detailed, extraordinarily still, wrapped in a beautiful sensory distortion. It only takes a light breeze to ruffle the snow and release trapped air, returning sounds to their ordinary volume.[2]

Snow may look dormant, but underneath the white blanket life barrels forward. Melting snow teems with sixty different strains of algae including the most common one, *Chlamydomonas nivalis*. This photosynthesizing microbe thrives in summer Arctic snowbanks and can tinge snow a vivid pink or red. The product of springtime mating, *C. nivalis* zygotes must survive the summer thaw before they can reactivate and keep growing in fresh winter snow. They protect themselves from blinding sunlight with a biochemical called astaxanthin, which absorbs UV radiation as well as heat and gives the snow its pink blush. *C. nivalis* also scents the snow with the faintly sweet freshness of watermelon—hence its nickname "watermelon snow."[3]

Next summer, split a chilled fresh watermelon, carve a slice, lift it to your nose, and inhale. A melting Arctic snowbank, marbled with pink streaks, smells the same way.

GINGER

It's warm and enveloping, this smell, but also vigorous. Smelling ginger welcomes you tenderly but also wakes you the hell up. The scent combines the spreading deep heat of

EXERCISE

Get a smell-training kit
(or assemble your own).

Those who suffer from smell disorders, whether total lack of smell (anosmia) or a distorted sense of smell (paranosmia), can sometimes regain their skills with smell-training kits. The concept is simple: every day you sniff a few different essential oils, observing and imprinting each smell in your memory. Ideally you'll notice a strengthening sense of recall as each day passes. This practice may also help you observe those smells "in the wild" when you encounter them.

Those with severe smell impairment may need to start slow, with only a few smells daily. But anyone can train their sense of smell to become more discerning. The same basic principle applies to how perfumers train: they systematically sniff their way through families of smells, working up to finer gradations between smells within groups. (The main difference is that perfumers are often memorizing specific smell molecules, versus smells as a layman might identify them.)

cinnamon with a bright citrus burst. It smells of sunshine, the tropics, the somnolent pleasure of lazing on a beach compelled to motion only by your body's inclinations and your mood.

It's unabashedly juicy. Fresh, unpeeled ginger smells like nothing, but remove the peel and it's instantly wet and fragrant. Ginger's hairy fibrousness dries out quickly but floods again with mysterious liquid stores when cut into. You refresh it simply by disrobing more of it.

Ginger is a durable rhizome, easily transported over oceans, that will grow in any tropical climate and is now consumed in almost every cuisine. Yet despite this ubiquity, ginger still retains an exotic feel, a quality of a special treat. In the early days of colonial exploration, ginger's scent was heavy with mystery and evocative of paradise to Westerners. It's a heterotopia, a slice of sublime pleasure grated into the everyday.

Ginger spells excitability in all its forms, most especially sex. There's a Venetian saying about the tempestuous character of redheads or "gingers": *Rosso de pelo, cento diavoli per cavelo,* "Redheaded, a hundred devils per hair." Redheads may not smell of actual ginger, but they are frequently associated with the spice's exquisite sensibility and wildness.

Across many cultures, ginger has been considered an all-around aphrodisiac capable of fixing any bedroom complaint. According to Galen's theory of humors, ginger was both hot *and* wet: its heat incited flagging libido and boosted sexual performance, while its wetness increased sperm count and fertility. In this wetness ginger differed from cloves, cinnamon, or—at the most extreme—pepper, whose dry heat fixed impotence and premature ejaculation and improved sexual stamina without elevating sperm count. Of course, humor theory was an elusive game of balance. Those whose lusts were allowed to run too high might dry themselves out so much that their hair grayed or fell out.[1] The corollary also applies: if your goal is to kill sexual desire or if your humors are already well balanced, eating too much ginger and other sexy spices can be too much of a good thing.

Here we turn to medieval Europe's premier sexologist, a Benedictine monk named

Constantine of Africa, whom we first encountered in the cinnamon section in chapter 2. He translated famous medical treatises from the Arabic, spreading the scientific writings of Avicenna and the Tunisian physician Ibn al-Jazzar among others across Europe.[2] Constantine's eleventh-century treatise *De coitu* (*On Sexual Intercourse*) includes many recipes using ginger to promote sexual healing.[3] For a nooner, Constantine liked a gentle concoction of cloves steeped in milk. For a marathon evening, he suggests a powerful nightcap of ginger, pepper, rocket, Spanish fly, sugar, and skinks (a kind of lizard).[4]

Ginger ranked foremost in Constantine's list of aphrodisiacs. Others included chickpeas, pine nuts, egg yolks, warm meats, brains, arugula, and testicles—from roosters to calves—as well as grapes, melons, and the phallic banana.[5] Eating all of these foods complemented drinking heavily spiced wines, a libido-booster at many celebrations but especially weddings. The idea was to encourage sexually inexperienced newlyweds with debauched guests taking the lead. This practice—prevalent among the well-heeled in Europe from Roman times through the Middle Ages—echoes visions of Muslim paradise as outlined in Sura 76 of the Koran, in which beautiful houris serve martyrs "cups brimming with ginger-flavored wine from the font of Selsabil."[6]

As a modern aphrodisiac, ginger has gone underground and niche. On the more extreme end of fetishes, "figging" is a BDSM practice in which a peeled ginger plug is inserted into the anus, an almost intolerable pain keenly pleasant to adherents to that sort of thing. More generally, a whiff of ginger is a welcome reminder that sex can cheer anyone up. Like ginger itself, it's a readily available indulgence that improves nearly every day (or dish). Take these lines from the satirical fifteenth-century French poet Francois Villon, who landed multiple times in prison for his misadventures sexual and otherwise. Villon bequeathed his torturer—called the "woodworker" in his text—with ginger in an effort to perk up his love life:

I give the woodworker
One hundred stems, beads and tails

Of Saracen ginger,
Not for coupling his boxes
But getting arses under the sheets,
Stitching sausages to thighs,
So the milk surges up to the tits
And the blood races down to the balls.[7]

ROSEMARY

It's a sprightly and pungent smell, more forest-filled than you expect. Bite into a rosemary leaf directly, and the taste and smell are astringent, cool-edged like mint. It doesn't seem edible, actually, and you briefly register the oddity of eating a plant so like pine trees in appearance and scent. (As it happens, the two plants are distant cousins.) Ancient cookery in the West didn't use rosemary as a culinary ingredient (although the ancient Mesopotamians ate it regularly).[1] Rather they had symbolic, higher-concept uses for this aromatic weed.

Rosemary's smell stirs the pot of memory and remembrance, and its own history goes deep. The ancient Romans burned rosemary sprigs as a cheaper alternative to incense.[2] Rosemary is also an ingredient in a four-thousand-year-old perfume unearthed by archeologists in Cyprus, blending lavender, bay, rosemary, pine, or coriander.[3]

Another of the world's oldest perfumes, Hungary Water, dates to the fourteenth century and included rosemary as a chief ingredient. Hungary Water was reformulated by perfume historians and can now be smelled in re-created form at the Osmothèque smell archive in Paris. The perfume's recipes vary depending on the source, but always feature rosemary flowers mixed with pennyroyal and marjoram. Later variations included bracing scents like citron, lavender, sage, and orris.[4]

Sometimes rosemary signaled remembrance of a literal kind, in terms of professional knowledge. Rosemary was core to an inscrutable ritual in the medieval French guild of *talemeliers*, the forerunner to bakers. After years of apprenticeship and paying his dues, an applicant to the rank of master baker brought a potted rosemary bush to the guild leader's house. The guild leader received him in a room reserved specifically for this ritual and inspected the plant's decoration—its branches were hung with candies, sugar peas, and oranges—as well as the plant's shape, smell, and roots. All of these served as symbolic proofs of the applicant's professional skills worthy of master status.[5]

Rosemary's scent mixes sex, romance, and death with wanton ease. Woven into crowns worn at ancient Roman funerary celebrations, rosemary enjoyed a long run associated with romance—mixed with sweet waters, it featured in many medieval bridal rituals. But its widespread use during the plague outbreaks of 1603, 1609, and 1625 conflated rosemary once again with death. Those seeking to fend off the plague armed themselves with strong, pleasant scents like rosemary (see the cannon fire section, chapter 4, for more). Wearing scented pomanders stuffed with rosemary, navigating around bonfires of burning rosemary designed to fumigate the plague-riddled streets— all these made rosemary's scent overshadow romance with intimations of mortality, at least for a generation.[6]

Which is not to say that death isn't slightly sexy; rosemary definitely benefited from *le petit mort*. In England, on the eve of St. Mary Magdalen (July 22) young virgin girls still dip rosemary in a mixture of rum, wine, gin, vinegar, and water, and then tuck the sprig between their breasts before going to sleep, in search of clairvoyant dreams.[7] Similarly, brides in Victorian England carried sprigs of rosemary on their wedding day to signify that they brought to a new life memories of the old one.[8]

PINE

It's brisk and invigorating, this smell. Its spicy warmth scrubs the nose and straightens the spine. It delivers the same trigeminal-nerve hit of camphor (see chapter 5) or lavender (see chapter 7). But pine's brio is gentler. Pine scent is warmer, woodier, more embracing than either of these. Even as this smell bucks you up, it also prompts you to slow down. It rinses you with a sense of relief.

Pine smell diffuses rapidly in open air but will fill a semi-enclosed space with wonderful intimacy. This smell is inseparable from the miniature-cathedral world of air a large pine tree encloses within itself: its arched bower of branches, the bushy gaps between needles, the tamped-down pads of dried needles blanketing the forest floor. It smells like a deep, protected repose.

But for all its bigness, this is also a shy smell. Wander outward into the world in the hopes of gathering pine scents—as I did for this chapter—and at first you'll meet with disappointment. The scent crops up everywhere when you're not looking for it and goes elusive when hunted directly. I gathered a collection of pine branches at a beach cabin in upper Michigan and was puzzled by their lack of pungency. I stepped under a Douglas fir to think through the conundrum, all the while breathing in its canopy of faintly scented, stirring air. And that's where I struck gold: clear yellow rivulets of resin ran down the tree trunk in front of me like tears. I wiped them onto a wet paper towel and sniffed that.

The fragrance was almost unbearably concentrated: simultaneously warming and cooling, a bracing green whiff, a shot of dappled sunshine.

That warm-cool embrace is telling. Even when sun-warmed—a pine tree shading a beach path in deep summer—a whiff of pine signals a place that also knows hard winter. Sooner than you might imagine, this pine grove will be again heavy with snow. The surrounding landscape will be radically altered by the seasons; only the pine trees themselves won't.

The smell of snow marks the golden, if temporary, stillness after snowfall. The smell of pine, while kindred, operates on a fundamentally different timescale. Pine trees are among the oldest living plants on earth, evergreen and unchanged by the passing of hundreds of seasons. The time spans they suggest are aeonic.

Together with bamboo and plum, pine trees complete a classic motif of Chinese art called the Three Friends of Winter (*Suihan Sanyou*), symbolizing the Confucian values of steadfastness, perseverance, and resilience. The Japanese word for pine, *matsu*, is fittingly a homophone for their word meaning "wait."[1] The ninth-century Chinese poet Bo Ju-yi wrote these lovely words about the old pine trees in his backyard: "They are 'useful friends' to me, and they fulfill my wish for 'conversations with wise men.'"[2]

I switch perfumes all the time. If I've been wearing one perfume for three months, I force myself to give it up, even if I still feel like wearing it, so whenever I smell it again it will always remind me of those three months. I never go back to wearing it again; it becomes part of my permanent smell collection.

—Andy Warhol, *The Philosophy of Andy Warhol*
(From A to B and Back Again)

It's the terpenes from pine resin that imparts most of the tree's scent. Pinene gives the smell its refrigerant, bracing effect and limonene its dash of citrus sunniness. Much like cannabis (see chapter 6) or pine's cousin plant cedar (see the section on freshly sharpened pencils, chapter 5), pines produce resin to repel insects. Conifers—a category which includes pines, firs, and spruce trees, among others—release even more terpenes on hot summer days. Clouds of water vapor form more easily around the terpene molecules, helping to block sunlight and thus cool the forest. In other words, the sense of an alternate world encircled by pine trees is real. Pine trees do create their own microclimate—shaded, contemplative, palpably different from the outer world.[3]

Breathing in pine-scented air feels life-giving in a way that's difficult to capture. Modern scientists have observed a kind of natural aromatherapy at work when we stroll through a forest. Plants emit aromatic compounds called phytoncides that may increase our production of white "disease-killer" blood cells, so walking through plant-rich environments saturates us with phytoncides with a possible immune-system boost.[4] The Japanese have coined a term for strolling through a forest aimlessly to improve one's well-being: *shinrin-yoku*, or "forest bathing." While any lovely forest will do, pine forests tend to be the most silent and most redolent. Pine trees darken the sky like skyscrapers, but they suggest with their towering trunks and spreading branches a civilization very different from ours.

Pine trees can restore health in more literal ways, too. Pine oil can relieve congestion and head colds or massage away aches and pains. Both pine bark and its needles contain a lot of vitamin C. Five hundred years ago, the French explorer Jacques Cartier's crew almost died outside Quebec from scurvy. A local Iroquois named Domagaia brewed them pine tea, from a plant the Iroquois called *Annedda,* and narrowly saved their lives. Its effect was swift and miraculous: after drinking the tea only two or three times and applying the tea dregs directly to their wounds, the sailors' skins cleared of wounds, their legs reversed their swelling and gained in strength, their gums ceased bleeding and loose teeth grew firm again. Domagaia had recently taken the remedy himself for the same symptoms.[5]

In a real sense, then, the hardest stretches of winter resemble a long maritime voyage. The length feels uncertain; stores of all kinds, whether material or emotional, run perilously low. The Iroquois knew pine trees—evergreen, somnolent, and life-giving—would carry them through to spring.

10 OTHERWORLDLY

Smells that defy easy classification. This chapter opens a vista of smell-making as a billion-dollar industry as well as a bridge traversing heaven and earth and the gaps between smells and language.

NEW BABY

EXTINCT FLOWERS

AN INVENTED SMELL

ECTOPLASM

THE ODOR OF SANCTITY

OLD BOOKS

NEW BABY

For me this smell reaches back eight years, to when my son, Lev, was born.

He smelled thrumming and warm, a tiny meticulous engine. The smell empha-sized his body's compactness but also enlarged his physical presence, like a nimbus or halo. In one big inhale, you could take in his entire scent of his little toes all the way to his head, tufted with curly dark-red hair. The notes I detected—fresh milk, a powdery dryness, smears of creams and unguents, crisp clean laundry not yet sodden—are the smells of care, the nonstop roundelay of keeping a little person alive and well. What a new baby really smells like is an otherwise impossible mixture: astonishment with contentment.

New-baby smell is memorable because it embodies an atmosphere both literal and temporally brief: newborn time, that fever dream of sleeplessness, jumbled non-activity, and quiet euphoria. It emanates from, and circulates entirely around, the newborn himself. You inhale this scent in enormous gulps, and it sustains you.

That notion isn't entirely fanciful. In a small study, a group of women including moth-ers and non-mothers were asked to smell a series of unidentified scents. When asked to sniff newborn scent (extracted from unlaundered pajamas), all the women's dopamine pathways lit up, the mothers' most of all. While the smell is less studied in fathers, all new parents know that new-baby smell delivers a powerful, and energizing, jolt of pleasure.[1]

Mothers and babies recognize each other first by smell. Newborns will doggedly navigate to the mother's breast, guided by scent alone.[2] With overwhelming accuracy, mothers can identify their own baby by smell within an hour of birth.[3]

Newborn smell is fleeting, evaporating within six weeks of birth. Some speculate the smell vanishes with the dissipation of vernix caseosa, the white cheese-like substance that covers a baby's body at birth. Or it might be the fading scent of amniotic fluid as the baby dries out from its watery origins and adjusts to living on land with us. In one small study, fathers and mothers were asked to sniff bottles of amniotic fluid and choose the one belonging to their child. Nearly all the fathers, and most of the mothers, chose correctly.[4]

Our bodies' smells change as we age. A 2020 study probed why babies smell so pleasant yet older children less so as they mature. Evidence indicates mothers can detect a child's developmental stage by their bodily odor, possibly tracking changes in the mother-child relationship. Babies and younger children smell sweetly, urging mothers to care for them during their most vulnerable stages. As children mature into teenagers, though, their bodily smell curdles and grows ranker, signaling the time for more separation and independence.[5] We're constantly signaling our age by smell, either attracting parental care or pushing it away.

Our bodies' smells keep changing into very old age. Consider the slightly grassy, greasy scent you might think of as "old people smell"—and the Japanese, more reverential of the elderly, call *keraishu*. This scent comes from elderly skin emitting higher concentrations of a compound called 2-nonenal. This shift may happen with changes in metabolism as we age, or due to fluctuations in the overall mix of chemicals our bodies emit. Levels of 2-nonenal do rise steadily as we get older, but "old people smell" is not usually detectable as such until a person celebrates their seventy-fifth birthday.[6]

Evolutionarily speaking, why should age be detectable by smell? One reason is that it might inform mating choices, although finding a younger mate isn't automatically better. Some animals prefer to mate with older animals who've proven their longevity, with all the genetic advantages this confers to offspring. In other words, the sex appeal of a "silver fox" isn't entirely irrational.

Time moves differently for very old people, newborns, and the caretakers who attend these two groups. And that strange time is imprinted in scent. I kept a diary of Lev's baby and toddler years; whenever I wrote in it always felt hurried and perfunctory. Some entries aren't written out at all, just notes jotted down. How could this dashed-off, highly interior nonsense capture any memories that would be legible to me later? But opening this diary now cracks open a vanished atmosphere. It's a time capsule permeated entirely by the smell and presence of a brand-new baby. Its redolence is intoxicating, its immediacy jarring.

I find the old saw about parenting is true: "The days are long, and the years are short." Whenever I paused to smell my baby, it pinned the moment down. It cleared an oasis of calm that never lasted long but would reemerge over and over unexpectedly, like a mirage you could briefly reach before it dissolved. It embodied in smell all the contradictory feelings about raising a child that unfurl in quiet moments: it's boring, it's thrilling, it's exhausting, it's creative, it's aggravating, it's all over in a twinkling and you marvel at its brevity and speed only belatedly. Whether we wish time to speed up or slow down is irrelevant. It merely continues. A baby's smell perfumes and punctuates a moment of stillness that's temporary. Now all elbows and knees, my eight-year-old clambers into and out of my lap for a hug that lasts only long enough for me to smell deeply into his hair or neck. Because he loves me, he pauses his nonstop motion and we bask briefly in each other's smells. *Ahh.* Here is the vanished newborn, here is the momentary respite of our connection. Then he pinwheels back into action, away from me again.

EXTINCT FLOWERS

I had planned to globe-trot the world and smell two extinct flowers for this chapter—but thanks to a global pandemic, those plans went up in smoke. One of the flowers is *Orbexilum stipulatum,* or Falls-of-the-Ohio Scurfpea, where I had expected to sniff it at the Philadelphia Museum of Art. The other is *Hibiscadelphus wilderianus Rock,* or *Maui hau kuahiwi*, "mountain hibiscus" in Hawaiian; when I made these plans, that smell was on long-term exhibit at the Wellcome Collection in London.[1]

How is it possible to smell an extinct flower? The science gets no less miraculous when you summarize it. Synthetic biologists at Ginkgo Bioworks began by visiting the Harvard Herbaria, a vast library of pressed plants and flowers, and snipped pinky-nail-sized clippings from quite a few of them. These clippings were pulverized, then examined for genes that produced enzymes called sesquiterpene synthases (SQS). These are the enzymes responsible for many flowers' scents. The biologists found a handful of these genes but needed to stitch the fragmentary sequences together into 1,700-character long strands. The scientists filled in the blanks by consulting the living DNA of similar plant species, hoping that their guesses would be close enough to make a functioning gene from the extinct plant material. By means of botanical guesswork, they cadged together more than two thousand such genes, then synthesized brand-new DNA from them using a DNA printer. The scientists then inserted the new, frankensteined SQS genes into living yeast cells in test tubes, hoping that the genes would reanimate in their live hosts and produce molecules that smell. Among the thousands of synthesized SQS genes they grafted onto live yeast, the genes of three extinct flowers sprang back to life, producing a very faint stream of their old scent molecules.[2]

This is an exceedingly expensive way to produce large amounts of smell molecules. So smell researcher and artist Sissel Tolaas took the list of molecules produced for each plant as a starting point and re-created the various plant smells by using the same or comparable smell molecules sourced from International Flavors & Fragrances, among other labs. These reconstituted smells are then pumped into a glazed diorama in an exhibit hall of a museum. Each installation created by the artist Alexandra Daisy Ginsberg re-creates not just each flower's smell but also includes boulders that used to surround the plants, a digital soundscape and a video reconstruction of the landscape at the time of the plant's extinction.[3]

That's the how. Now for *why* you'd try to accomplish this. A technologist might answer: ancient DNA is potentially useful code that's currently lost to us. But that

EXERCISE

Verbalize smells.

Describe whatever you smell in words, however crudely or approximately. Just verbalizing what we smell can boost your smell capabilities.

Olfactory receptors—the receptors in our noses whose proteins bind to smell molecules—regenerate every four to eight weeks and change in response to whatever new smells they encounter. Smelling new scents, articulating what you smell in words, learning to identify similar smells—in short, practicing your sense of smell is brain-building, particularly in older adults.*

Treat smells like puzzles or word games, a sieve through which you can scramble and refract the world. One scholar, writing about the many smells in literature, put it nicely: "The investigation [of smell] is the investigation of everything else."

---

* https://www.sciencedaily.com/releases/2020/11/201130131517.htm.

argument, while noble, is also pretty abstract. A more compelling reason is this: re-creating lost scents makes extinction seem realer. It stabs us in the heart. In this time of mass extinctions, no one can mourn every single species' passing. The losses mount to an incalculably high number; it's numbing. But every lost species also means a lost habitat full of symbiotically related species, themselves probably vanished, too. One extinct flower, sniffed in the present, conjures an entire lost world.

A tall, slender plant with pale-white flowers packed into small globes, *Orbexilum stipulatum* was last seen in 1881 on Rock Island near Louisville, Kentucky—the only place on earth it seemed to have grown. Rich in Devonian limestone, this island was situated near a two-mile stretch of rapids in the Ohio River. In the 1920s, Louisville built U.S. Dam No. 41 and an associated hydroelectric plant, which flooded Rock Island and destroyed this plant's habitat—and with it, all traces of *Orbexilum stipulatum*.[4] *Scientific American* describes this flower's smell as "woody, peppery and balsamic." At the time when U.S. Dam No. 41 was built, three of my grandparents—now all dead—were children growing up near those same banks of the Ohio.

Hibiscadelphus wilderianus exudes a more complex smell, evoking a similarly vivid lost landscape. *H. wilderianus* grew only on dry, well-drained lava fields near Maui's volcanoes. The last living plant was spotted in 1912 on the slope of Maui's Mount Haleakalā. At fifteen feet tall, its hibiscus-style flowers unfurled to attract the Maui honeycreeper—nectar-eating songbirds that had been *H. wilderianus*'s best pollinators until the birds themselves went extinct in the nineteenth century. You can imagine the scene as observed by botanist Gerrit Wilder (for whom the plant is named): a vast sweep of lava fields, which became gradually overrun by ranched cattle brought in by European settlers. The cows rubbed the mountain hibiscus' bark raw, while rats ate its remaining seeds. Wilder rescued the last living plant he found and nursed it into blooming in captivity—a melancholic and fleeting success. *Scientific American* describes its scent as a "woody core of pine or juniper," with "flashes of thyme and citrus" and a "smoky hint of sulphurous dirt" lurking in its background.

So many ironies circle around these smells for me. They started as wholly imaginary, an intriguing thought experiment, but became realer as I booked the flights and made concrete plans to smell them myself. When COVID-19 hit, all those plans unraveled and the smells returned to the imaginary. Yet they seemed all the realer for that fleeting brush of contact, if only in my head. Someday soon, I hope.

AN INVENTED SMELL

Spritzing from the spray bottle, I sniff a silvery, unreal arc that seems almost digitized—although perhaps that's my imagination playing tricks on me. Blended into my wrist, the scent warms up and becomes more supple and analogue. It's woodsy and peppery, glinting with amber and gold tones. There's a floral creaminess that gives the scent body, yet it's also clear and buoyant. This smell is secretive but not shy. It invites you to draw closer and then maddens you with its elusive, low-key charm.

This is Iso E Super, one of the more famous smells invented by chemists in a lab—specifically by two Americans, John B. Hall and James M. Sanders, working for leading perfume and flavor manufacturer International Flavors & Fragrances (IFF). Iso E Super—short for Isocyclemone E—was patented in 1973 and has subsequently appeared, at increasingly high concentrations, in many iconic perfumes. I am sniffing the logical endpoint of the craze for this molecule: Molecule 01, a perfume consisting 100 percent of Iso E Super and the brainchild of experimental perfumer Geza Schön.[1]

Iso E Super isn't just a synthetic copy of a preexisting smell. This molecule derives from no real-world object other than itself. Before its invention, it didn't exist in nature. Iso E Super and smells like it illustrate a very weird fact: unlike colors, we *can* invent brand-new smells.[2]

The maximum number of smells humans can identify isn't fully known. (The highest estimate so far, debated but not outright disproven, reaches a trillion.)[3] Our noses are equipped to smell many, many molecules that we haven't yet encountered—or that may not even exist yet.

The business of inventing smells starts with the desire to capture smells in nature. From this urge sprang all the perfumers' traditional arts: essential oil pressings, enfleurage, myriad distillation methods, solvent extractions. (For more on this, see the jasmine section, chapter 1.)

Capturing smells eventually led to synthesizing them in a lab. Cinnamaldehyde—the dominant smell-note in cinnamon (see chapter 2)—was among the first synthesized smell molecules in 1834.[4] First the goal was simply to copy smells from nature, but into copying crept the impulse to perfect them. Distillation can distort the original scent or fail to capture its full charm. Distillation means heating and cooking the flowers, which not only changes their smells but can completely destroy them. That explains why we don't have essential oils of lily of the valley or jasmine; they're too delicate to distill.

Synthesizing smells can eliminate these distortions, edit out unpleasant notes from the original smells, or foreground the desired smell molecules. Synthetic smells are often cheaper than "natural" ones, yet molecularly identical. Synthesizing is also environmentally friendly: it protects natural smells from getting hunted into extinction. Synthetics sometimes get a bad rap, but in many ways synthetics improve upon naturally derived ingredients: they boost a fragrance's longevity and help it develop more beautifully as you wear it.[5]

The hunt for new smells, whether preexistent or invented, is relentless. Right now researchers are plunging deep into remote jungles, looking for exotic plants whose smells might be interesting to capture and synthesize for use in perfumes. Capturing a rare plant's smell need not disturb the plant one bit. "Headspace" technology requires surrounding the flower with a glass globe that slowly fills with the various scent molecules the plant exudes over a twenty-four-hour period.

The scent-saturated air inside this globe is rushed to a laboratory, where researchers use gas chromatography–mass spectrometry (GC-MS) to reverse-engineer the smell molecules present in it. Equipped with this technology and bankrolled by global smell- and flavor-manufacturers, fragrance chemists are hunting brand-new smells, racing against both accelerating mass extinction and each other.[6]

Back to Iso E Super, the invented smell slowly opening up on my wrists. It's the product of a chemical reaction between myrcene (a terpene found in nature; see the cannabis section, chapter 6) and 3-methylpent-3-en-2-one. It earned the "Super" designation after IFF chemists tweaked the original compound to improve its smell qualities; it has since been tweaked many more times. For a time it was a so-called captive smell, protected by patent exclusively for use in IFF's perfumes. (For more about smells, patents, and trademarks, see the Play-Doh section, chapter 8.) A capsule history of Iso E Super is an object lesson in how an invented smell molecule's story plays out if it's commercially successful.

Iso E Super appeared in some functional perfumes—the smells added to laundry detergents and household cleaning products—and finally broke into the fine-perfume market with Halston Woman 1975. The molecule formed only a miniscule part of this perfume's formula, but it paved the way for more perfumes—many blockbuster hits— to incorporate Iso E Super in increasingly higher doses. This roster includes Christian Dior Fahrenheit (1988, 25 percent Iso E Super), Lancôme Trésor (1990, 18 percent Iso E Super), and Féminité du Bois (originally Shisheido, now Serge Lutens, 1992, 43 percent Iso E Super). Until 1995, when the Iso E Super patent expired, IFF had the exclusive right to this molecule, giving them a sweet competitive advantage as they formulated perfumes for iconic houses like Christian Dior and Lancôme. (IFF, Firmenich, Givaudan, Takasago, and Symrise are the chemical wizards who actually create all the big-name perfumes; they're simply works for hire that the fashion houses brand and market to consumers.)

In 1995 the floodgates opened: perfume manufacturers other than IFF could use the original Iso E Super molecule as well as tinker more freely with its chemical composition. In the mid-1980s, analysts at Givaudan had already reverse-engineered the Iso E Super molecule to study its properties more closely. (This practice, however chagrin-inducing to a smell's inventors, is common among fragrance chemists and perfectly legal.) To their surprise, the Givaudan analysts learned that pure Iso E Super hardly smells of anything. What smell qualities it possesses are due to a component that's only 2 to 5 percent of the commercial product but packs an outsized wallop. The Givaudan team isolated this submolecule, named it Iso E Super Plus, and then raced to patent it themselves. In 2005 Geza Schön launched Molecule 01, which I'm wearing now. Further tinkering with Iso E Super and its offspring molecules will continue, just as the chase for fresh new molecules will.

Inventing a smell sounds thrillingly blue-sky—and, in the early stages, it is. But reality always intrudes, hemming the molecule in from multiple sides. Iso E Super Plus was a concentrated fragrance bomb, but it was too expensive to incorporate into mass-market perfumes. Invented smell molecules must not only clear the hurdle of affordability, they also need to pass tests for health and safety, potential allergic reactions, biodegradability, and eco-friendliness. Molecules are constantly getting yanked from international lists of ingredients acceptable in perfumes—which suggests the perfumers' mania to find new smells isn't just paranoid. It's also practical.[7]

SNIFFING MY ISO E SUPER AGAIN, I'm struck by its intimate, unfolding quality. Not particularly stiff to begin with, it unbends over time and becomes looser. It calls to mind something I read about Lancôme Trésor, a perfume combining Iso E Super with Hedione—another famous molecule invented in a lab—with galaxolide and α-methyl ionone. Perfumer Sophia Grojsman, who created Trésor, gave the fragrance a working title: "Hug Me." The combination of these four smell notes—called an "accord"—is now known as the "Hug Me accord" or the "Grojsman accord."

Molecule 01 smells like that: the scent you find by burying your nose into your lover's neck as you embrace them. (For more on this, see the skin section, chapter 6.) It's insouciant and therefore sexy. It's a singular idea that's survived organic chemistry, industrial manufacture, a barrage of regulatory tests—all to nestle into your skin as if it's known you always.

New smells await everyone, at every age. Given some taste for adventure, even very old people can encounter brand-new smells they've never sniffed before. What other sense offers such ample white space?

ECTOPLASM

To make my own ectoplasm, I mixed up egg whites, glue, and soap shavings into a paste and then soaked cheesecloth in it. Warmed up to room temperature, its scent was exceedingly faint: a whiff of wet freshness, a suggestion of food, a barely detectable shift in atmosphere like a change in weather. It's no wonder its smell was compared very often to ozone.

What is ectoplasm, and what significance was historically attached to its smell? It all starts with the spiritualism movement—one might fairly call it a craze—that ran from the 1870s into the early twentieth century. Its chief ritual was the séance, in which the living could communicate with the spirit realm of the dead. A female medium—sometimes veiled or blindfolded, sometimes partially or totally naked—would shut herself up in a "spirit cabinet" with the lights dimmed and summon spirits for the benefit of an onlooking crowd. Groaning as the trance took hold, ectoplasm might begin to dribble from the medium's bodily orifices—mouth, ears, eyes, navel, vagina, or rectum. This phenomenon was known as "materialization" and provided palpable evidence of the spirit realm penetrating our own.[1]

Believers regarded ectoplasm as a quasi-living substance. Mediums described ectoplasm's initial appearance like a luminous vapor effervescing into a steamy substance. Sometimes, they claimed, it felt like "fine threads being drawn out of the pores of the skin."[2] Once it had fully emerged, ectoplasm was usually a white, "viscous substance . . . from which spirits make themselves visible forms . . . alive, sensitive to touch and light . . . cold to the touch, slightly luminous and having a characteristic smell" of ozone, that steamy, metallic scent that presages a coming thunderstorm. In the 1920s, a French doctor who studied ectoplasm in a series of experiments describes its creepy self-animation: "The substance is mobile. At one moment it evolves slowly, rises, falls, wanders over the medium, her shoulders, her breast, her knees, with a creeping motion that recalls that of a reptile . . ." Ectoplasm's smell is emphasized in many contemporary descriptions, clinching its reality.

Ectoplasm was a flexible stuff that the spirits used to manifest themselves physically during the séance. Sometimes a spirit would, via ectoplasm, morph into a "pseudopod"— amorphous shapes like limbs or webs, suggesting a teeming presence of spirits pressing into our world. Or ectoplasm might form into a spirit's body or just their hand or foot, like a plaster cast floating in air. Other times ectoplasm might become a scrim upon which the spirit projected its face—a kind of death mask called an "ideoplast."

This all sounds ridiculous and definitely was. Yet spiritualist materialization also interested credible scientists of the day, because ectoplasm dovetailed with several scientific inquiries happening during the same period. For nearly a century biologists had wondered about the secrets of protoplasm, a gelatinous substance in plant and animal cells which, they thought, might form the basis for all life. They speculated: could ectoplasm be protoplasm's uncanny cousin?[3] In the 1870s, Sir William Crookes, a prominent chemist and spiritualist supporter, worked with Michael Faraday to investigate a new form of matter they termed "radiant matter." Sir Oliver Lodge—a pioneering physicist whose work in radiotelegraphy paved the way to the theory of relativity—didn't hesitate to expound on the nature of ether in a 1925 book. He thought of Ether, a term he capitalized, as a kind of super-matter connecting mind, the earthly world, and the spiritual beyond. Ether, according to Lodge, transmitted vibrations from one bit of matter to another. Since its frequency differed from ordinary matter, he theorized, vibrating ether *might* materialize briefly as ectoplasm.

In 1882, a group of scientists and scholars founded the Society for Psychical Research. At one event a decade after the Society's founding, many of its members gathered at the home of Dr. Charles Richet, a Nobel Prize–winning scientist who originally coined the term *ectoplasm*. Their purpose in gathering was to observe the medium Eusapia Paladino's work at close range. According to written accounts by participants, floating guitars played themselves, bells rang spontaneously, vasefuls of jonquils appeared, and ectoplasm oozed phantasmagorically over every surface.

Actual ghosts, permeable realms, brand-new forms of matter, the thrilling cusp of scientific discovery, witnessing it all live complete with actual, irrefutable smells. The overheated excitement around ectoplasm is still palpable.

Ectoplasm was irresistible *because* it was material. In the brief moments it was present, you could photograph it or study its properties: its texture, movement, temperature, and smell. It fluttered a veil hiding the unknown, through which an actual perfume wafted. Even if the ectoplasm itself disappeared, its fragrance still lingered.

What was ectoplasm really, and what did that smell like? My ectoplasm re-created the basic recipe in broad strokes but fell short in a few ways. Traditionally it contained gelatin from ground-up horse hooves, imparting a gamey whiff of barnyard and paddock. I couldn't find old-school gelatin of this kind and had to settle for citrus pectin. Adding the pectin did pump up the egg-sulfur smell of my ectoplasm robustly, if perhaps inaccurately.

What would really amp up ectoplasm's scent, I suspect, would be to smell it in situ. Before the séance, the medium might stuff ectoplasm into her orifices or hide it amid folds of heavy clothing; assistants sometimes helped her produce it at the opportune moment. One medium was known for materializing only while heavily veiled; ectoplasm would dribble out of her mouth or ears, through the veil.[4] Probably ectoplasm acquired some of its smell from the mediums' own bodies: their perspiration, saliva, or other secretions.

From smelling ordinary soaked cheesecloth to smelling ectoplasm required imaginative leaps. I'm missing several layers of that alchemy: the medium's warm skin, guttering candle wax, the anxious sweat of believers craning forward in rapt expectation. The astonishment of belief confirmed, with a lingering scent as proof.

THE ODOR OF SANCTITY

It is always sweet, this smell, and it's always astonishing. The scent itself can be pleasantly humdrum—baking bread, smoldering incense, fresh-cut flowers—but its emanation from an unperfumed body cannot be explained by mere facts. It resists logic, deliriously so. Smelling the odor of sanctity, one gets a touch lightheaded. After all, a whiff of heaven is threading up your nostrils.

This is the "odor of sanctity," various sweet smells that supposedly encircle certain holy Catholics during their lives and often after their deaths. The history of this concept begins with martyrs, Christians wreathed in beautiful smells as they were marched to the stake to die for their faith. Martyrs were often unperturbed by death and unkillable by the infidels (at least at first). When they did finally die, an unreally gorgeous perfume wafted over the scene, a sign that God was present to sweep the martyr up into paradise.[1]

Generally pleasant, the odor of sanctity sometimes shifted as action in the holy person's story unfolded. Take St. Polycarp, for instance, an early Christian martyr who refused to renounce his faith and was burned at the stake by Romans in Smyrna. One source describes the evolving smells as Polycarp died:

> The fire shaped itself into the form of an arch, like the sail of a ship when filled with the wind, and formed a circle around the body of the martyr. Inside it, he looked not like flesh that is burnt, but like bread that is baked, or gold and silver glowing in a furnace. And we smelt a sweet scent, like frankincense or some such precious spices.[2]

Like many forms of mysticism, the odor of sanctity got increasingly baroque over time. At first only martyrs exuded this scent, but later it was observed in Christian saints who hadn't died a martyr's death. With that shift came a shift in the smell's meaning:

now the sweet smell could indicate a living saint, a human being briefly dwelling on earth but already consecrated to heaven. Saints smelled literally otherworldly, and often that scent intensified upon their death and persisted long afterward.

Teresa of Avila is perhaps the best-known saint in this category. Known for her mystical raptures—described in her writings in very smell-forward terms—on her deathbed Teresa's body emanated a powerful fragrance of lilies and jasmine that pervaded her rooms and persisted for months after her death. Her fragrant corpse could also perform olfactory miracles. Sniffing Teresa's incorruptible corpse, or kissing her foot, was sufficient to cure pilgrims of anosmia, the inability to smell.[3]

Some saints' corpses or their sarcophagi oozed scented oils, which pilgrims could collect into little flasks. This oil, which often conferred healing powers, is called myroblysia, and the saints who oozed it are called myroblytes. A roll call of myroblytes includes a few big-name saints—St. John the Evangelist and St. William Archbishop of York—as well as many lesser-known saints with names like St. Walpurga, St. Menas, and St. Humilitas.

Sometimes the odor of sanctity would mingle with reeks, contrasting the mundane world with the divine one. Take Simeon Stylites, a fifth-century Syrian monk given to acts of extreme self-flagellation as penance for his sins. Once Simeon bound a rope around his waist tight enough to cause bleeding, then left it in place for a year. His flesh, rotting around the open wound, stunk so much that he was expelled from his monastic community.

The next phase of Simeon's self-mortification involved living alone atop a pillar, or "stylite," and inflicting various kinds of suffering upon himself there. This phase supposedly lasted thirty-seven years.[4] Toward the end of that period, legend has it that a king came to visit Simeon and behold this living, reeking monument to penance. A maggot fell from one of Simeon's gangrenous feet and landed at the king's feet. When the king picked up the maggot, it had transformed into a pearl.[5] Simeon's mortified flesh stunk more and more intensely until he died, but once he did expire Simeon's corpse exuded a glorious fragrance that intensified right up to his burial. The moral in smells is pretty clear: Simeon's earthly body reeked of human corruptibility, and the sweet deathly perfume proved his redemption through God's grace.

Another saint mixing foul and fragrant smells is St. Lydwine, a Dutch saint living at the turn of the fifteenth century. A serious invalid from her youth, St. Lydwine suffered from constant vomiting and bleeding as well as putrefied tumors. Despite constant disease, her rooms and body were always delicately perfumed. The smell shifted over time, too. At first Lydwine's rooms smelled like baking spices: ginger, cloves, and cinnamon. Later the scent refined itself into a bower of fresh-cut flowers: roses, violets, lilies. St. Lydwine's distinctive scent also had religious powers: it could penetrate visitors' consciences and even horrify demons.[6]

The scented body is surrounded by angels.

—Arabic proverb

Other religions besides Catholicism equate pleasant smells with moral goodness and bad smells with evil. When the Buddha spoke, his words were said to perfume the air around him. Bodhisattvas now listen to the Buddha's sweet words symbolically by burning incense, focusing on the scents to better concentrate on their prayers. (For more on *kōdō* and its ritualized use of incense, see the oud section, chapter 5.)

The scent of apples recurs a lot in stories from Shia Islam. In one tale, the angel Gabriel gave the Prophet Muhammad an apple from paradise. Muhammad bit into it, and its juices impregnated his wife Khadija with their daughter Fatima. Muhammad believed he smelled the fragrance of paradise in his daughter. Often compared to the Virgin Mary, Fatima married Ali and gave birth to sons Hasan and Hussein. The latter led a revolt against an oppressive Umayyad caliph in the seventh century CE and was martyred at Karbala; he is revered by Shia Muslims as the third imam and a hero.[7]

As children Hasan and Hussein received an apple, a pomegranate, and a quince from the angel Gabriel. The fruits were magical in that every bite one took was immediately replenished. The quince and pomegranate disappeared when Hasan and Fatima died, at which point the apple's special powers faded, too. Hussein carried this now-finite apple into his final battle at Karbala and bit into it just before he died, attempting to quench his thirst. The apple disappeared as Hussein expired, but the scent of apples later wafted from Hussein's tomb. Subsequent Shia martyrs like Qazi Nuru'llah in the sixteenth century also smelled like apples after their deaths, alluding to the foremost of Shia martyrs, Hussein.[8] Throughout all these stories, the apple smell connects a clear and God-given lineage of Shia notables.

Is the odor of sanctity bunkum? Strictly speaking, yes, probably. Science has partly explained away many of these smells. Quite a few Christian saints wasted away in starvation, so their bodies may have smelled sweet because they were undergoing ketosis. In a more fraudulent vein, at least one font of blessed myroblyte oil proved to be only vegetable oil.[9]

And yet. The tale of a martyr, punctuated with the right smells, could straighten any skeptic's spine—smells underscore reality with a kind of inarguable force. A beautiful smell occurring out of nowhere always snaps people to attention. It punctuates a moment, suggests a greater presence. It hints at realms much larger than our own—and then lets us actually sniff them.

OLD BOOKS

Inside its dry and musty hull, this smell contains great distances. You sense time travel, of course, but also the soaring aeronautics of ideas. Sometimes a great book sticks the landing, very often they don't. And sometimes books fail to jump high enough.

In smelling old books, you can smell actual geographic distances, too. Imagine all the cardboard boxes necessary to crate up a roomful of books, the slow trundle of cargo to a new destination, the books aging along with their owner from move to move. Readers who find this smell intoxicating are ruefully aware of how insane, how flatly contradictory of convenience, loving this smell can be.

From an old book filters up a whiff of dissolution and conjuring. Ghosts, like smells, are composed of particles, a levitating cloud forming a slightly denser shape than the surrounding air. The disembodied words of writers, whether dead or otherwise absent, find body in these volumes.

The litany of things book-smell resembles suggests elements of peaceful civilization: Well-rubbed wooden furniture. Leather bookmarks. Just-opened tobacco tins. Toasted almonds mixed with vanilla. Tea. Pressed flowers. Radiator heating pinging on. Candles and spent matches. Undisturbed dust, with its suggestion of perfect stillness and the happy dilation of hours spent reading.

This is an intimate smell that's not necessarily miniature. You can smell a single old book, riffle through its pages rapidly and let your nose bask in the scented breeze. But just as often you encounter this smell as a solid wall in a used bookstore or library. Writ large like this, old books smell like a constructed forest: ancient and druidical, exhaling to make their own atmosphere, a forgotten primordial home.

The smell of old books stems from their slow chemical decomposition. Books are largely paper, and paper is largely plants. But the materials from which books are made have shifted over the centuries—and those shifts, in turn, have influenced how different generations of books smell. Before 1845, paper in books was manufactured from cotton and linen rags. These plants contain a high percentage of cellulose, a kind of polymer that gives plants stiffness. Cellulose is largely stable in paper over the years unlike its cousin lignin, another plant polymer frequently present in paper.

The 1840s saw the invention of a process to manufacture paper from wood pulp fibers. This produced paper that was less durable than paper made of cotton or linen, but more plentiful and therefore cheaper. Wood-pulp paper has heavy lignin content unless you extract the lignin deliberately finer-quality papers tend to contain less lignin than newspaper. In trees, lignin binds cellulose fibers together and gives wood its extra backbone—which would seem like a desirable quality for bookbinding. What early bookmakers didn't realize was a lesson only learned over time: lignin is strong during a plant's life, but it's prone to wild and destructive decay after death. As lignin oxidizes, it breaks down into acids that eat away at paper's cellulose and turn the pages yellow. All of these processes of decay throw off volatile organic compounds, or VOCs, that impart a distinctive smell to aging books. Around 1970 acid-free paper was introduced to bookmaking, removing most of that calamitous lignin so that modern books degrade much more slowly. Even without lignin, however, cellulose eventually decays and releases scents.[1]

The smell of books fuses with the smell of rooms, and the readers who habitually occupy them. Books' fragrance reflects the environment they've long resided in: if that space is particularly humid, smoke-filled, sunny, or not climate-controlled. If those rooms happen to be iconic, like the Wren Library at St. Paul's Cathedral in London, then the smell itself can be considered a tangible part of a site's heritage. Analyzing the VOCs wafting from books, researchers can discern what materials the book was made of, pinpoint its age, and whether any disastrous processes are at work that preservationists might want to halt or reverse.[2]

Brand-new books also have a distinct smell, if a less singular one. Modern paper manufacturing uses chemicals like sodium hydroxide (or "caustic soda") to boost pH levels and swell up the pulp fibers. Those fibers can also be bleached in hydrogen peroxide, then mixed with copious amounts of water laced with alkyl ketene dimer, a sizing agent that makes the paper more water-resistant. These chemicals decompose at various rates, throwing off smells in turn. The various inks and adhesives used to bind books together do the same. As soon as a book or any object is made, it starts to unmake itself—a chemical process we can smell.

CONCLUSION

BEAUTIFUL-SMELLING OLD BOOKS, WITH THE BILLIONS OF WORDS INSIDE them, belie how few words exist in English to describe smells specifically. We usually settle for words that compare the smell with the thing emitting it, like *sugary* or *smoky*. Psychologists refer to the olfactory verbal gap—our inability to describe smells with words—and the related "tip of the nose" phenomenon, in which we recognize a smell but, frustratingly, can't name it outright. These phenomena seem universally human but may be more acculturated than we realize—the product of a modern industrial society indifferent to smells and out-of-practice in naming and describing them.

With every chapter of this book, I sniffed something and then attempted to capture that smell in words. At first the effort made me incredibly tongue-tied. The urge to blurt out IT SMELLS LIKE CINNAMON was overpowering. I had to learn to sit longer with the sensation of a smell, approach it from unexpected angles, experiment with metaphors and draw on other sense-modalities. Food smells were particularly encouraging: I already felt comfortable articulating what I taste, so learning that flavor was a secret back door into smell helped me believe I could overcome my own olfactory verbal gap. I learned to compare variants of the same smell to observe microtonal differences between them. I meditated on how each smell made me feel, what memories that smell triggered for me, and how the new facts I'd learned changed my thinking about that scent. All it took was patience, curiosity, and practice.[1]

Writing this book brought me back over and over to Wittgenstein's famous remark: "The limits of language are the limits of my world." In learning to smell more, I've stretched my world in a tiny but substantial way. As I type this sentence, the book exists only as a digital file on a computer. It's made of words, not actual smells, and it is (of course) odorless. Eventually the published book will take on physical shape. I wonder what it will finally smell like.

Not all languages are as impoverished for smell-terms as English is, or as out of practice in describing smells as English speakers are. Case in point are the Jahai and Semaq Beri people of the Malay peninsula, which psycholinguistics researchers have studied most intently. Both tiny hunter-gatherer tribes, the Jahai have only a thousand members currently and the Semaq Beri twice as many. Both groups live and hunt traditionally in tropical rainforests, relying on their noses to parse the smell-rich environment they navigate daily. Jahai speakers also use smells to organize social interactions and keep the world orderly. They are careful not to cook certain game animals on the same fire because their distinct smells shouldn't mix. Similarly they forbid brothers and sisters from sitting too close to each other lest their bodily smells mix in a kind of airborne incest. Jahai speakers monitor and manage the smells actively around them, so it's no surprise they have developed a nuanced vocabulary for this sense register. Learning a few untranslatable smell-words in Jahai opens up a vivid, unfamiliar world:

> For example, haʔɛ̃t (pronounced something like "huh-ed") is the Jahai word for the unpleasant smell common to tiger, shrimp paste, sap of rubber tree, rotten meat, feces, musk gland of deer, wild pig, burnt hair, old sweat, and lighter gas. It's a distinct smell from cŋɛs ("chung-ess"), which refers to the smell of petrol,

smoke and bat droppings—a stinging sort of smell. And distinct again from pʔus ("puh-oos") smells associated with old dwellings, some species of mushroom, cooked cabbage, or stale food.[2]

We English speakers don't have many smell-specific words, but the ones we have are promising. Pungent. Fragrant. Musty. Redolent. Aromatic. Malodorous. We should invent more.

FREQUENTLY ASKED QUESTIONS

. . . and where to find the answers in this book.

Why do old people's bodies have a distinctive smell?
New Baby, chapter 10.

Is there a universally loved smell?
Vanilla, chapter 2.

A universal stink bomb?
Durian, chapter 3.

How are perfumes made?
Jasmine, chapter 1, and An Invented Smell, chapter 10.

Which sudden shifts in body odor provide clues about your health?
Skin, chapter 6.

What's the difference between a pheromone and a smell?

Skin, chapter 6.

What does the world's oldest perfume smell like?

Rosemary, chapter 9.

How do other animals use smell to navigate or tell time?

Musk, chapter 6.

How do deodorants actually work?

Skin, chapter 6.

Can you trademark a smell?

Play-Doh, chapter 8.

Can you invent a brand-new smell?

An Invented Smell, chapter 10.

Is it possible to smell an extinct flower?

Extinct Flowers, chapter 10.

Do other languages have words for smells that English doesn't?

Old Books, chapter 10.

Which emotions can humans smell?

Skin, chapter 6.

ACKNOWLEDGMENTS

I first started thinking about smell as a book topic after our dear friends Heidi Enzian and Thomas Richter visited an exhibit of Smeller 2.0 and told us all about it. I won't replay the train of thought this sparked, which is already covered in this book's introduction. But I do want to give Heidi and Thomas a big herzliche shout-out. Their brand of curiosity is exemplary: they're both willing to entertain oddball questions and keen-witted enough to hunt down the answers. Those two are some of the smartest, most sensitive observers and people and I know—and I love them both dearly for it.

The project took on additional steam when I pitched Smeller 2.0 and smell-o-vision as a magazine article to Ian Bogost of *The Atlantic*. As it turns out, Ian and his editor didn't bite—but we did engage in a productive exchange that got my wheels turning further. Thanks, Ian, for helping me recognize how invested I could get in this topic.

I also enjoyed some lovely outdoor drinking and conversation in Berlin with Wolfgang Georgsdorf, the inventor of Smeller 2.0, and Alanna Lynch and Mareike Bode, cofounders of Scent Club Berlin. In true spontaneous-Berlin style, artist and smell researcher Sissel Tolaas invited me to a viewing of one of her scent-art installations based on her historical research of smells permeating a particular street in Berlin. These meetings all primed my enthusiasm about smell and provided useful starting points for further reading and research.

Sources are always amazing. To paraphrase a tweet I committed to memory, academics in particular will work extremely hard to get facts right for no other compensation than the pursuit of accuracy. That praise applies amply to this group of experts, academic and otherwise. Olfaction experts including Jay Gottfried and Michael Stoddart helped me shape my reading list and made sure the scientific inferences I drew were accurate. I took a beginner's perfume-making course at the Institute of Art & Olfaction, which led to a brisk email exchange with its founder and executive director, Saskia Wilson-Brown. Sommelier Alex Ring gave invaluable input as I wrote the wine chapter, as did wine instructor (and my father-in-law) Howard Spiegelman. Peter DiCola read the Play-Doh chapter with a lens on intellectual-property law (and a generous sense of humor). Steve Morrison described his personal smell loss and recovery of same through treatment. Adrienne Beuse and Nabil Valiulla are ambergris dealers who independently sent me complimentary pebbles of ambergris to sniff myself. Christopher Kemp, who wrote a great book on ambergris, put me in touch with them as well as provided input on my ambergris chapter. Experimental perfumer Geza Schön provided input for the invented smell chapter. Mikey Manning, chief brewer at Marz Community Brewing in Chicago, answered round after round of emails patiently and expertly about brewing beer. My sister-in-law Hannah Spiegelman is an ice cream expert who supplied recipes for both vanilla ice cream and durian sherbet. Other experts who reviewed various chapters for accuracy include Christina Agapakis, Tim Ecott, Iain Gately, Alexandra Daisy Ginsberg, Ryan Jacobs, Bronwen and Francis Percival, Patricia Rain, Jonathan Reinarz, Helen Saberi, and Chuck Stanion.

Clearing space and time to write is crucial. I've had several writer's residencies that have helped this project along enormously, including Hambidge Center for Creative Arts and Sciences; Virginia Center for the Creative Arts (VCCA); Write On, Door

County; and the Ragdale Foundation. I am also indebted to the dearly departed Writers Workspace of Chicago, founded by the indefatigable Amy Davis.

My editorial team is *the best*. My agent, Jen Carlson, first encouraged this project over Chinese food in Minneapolis and has been a fantastic advocate for the book as it took shape. Meg Leder, Annika Karody, Travis DeShong, Amy Sun, Christina Caruccio, Eric Wechter, Kristina Fazzalaro and the whole Penguin Random House team: what a delight to bring this book to life with you all!

Writing a book during a pandemic is no joke and truly a community effort. My friend Andaiye Taylor helped me shape my thinking for two chapters, line-dried laundry and cannabis. Other friends —Susy Bielak, Justin Lee Mark Caulley, Ezra Friedman, Theaster Gates, Seth and Jules Kim-Cohen, Anna Kornbluh, Ann Leamon, Adam Marks, Alice Mattison, Sianne Ngai, Hallie Palladino, Ian Quinn, Josh Rutner, Fred Schmalz, Hans Thomalla and Jina Valentine—listened to me talk about smells *a lot* and encouraged my enthusiasms at every turn. My parents, Mike and Mary Stewart, and my brother's family—Andrew, Danielle, Emily, and Maggie Stewart—all gave support and enthusiasm warm enough to radiate across socially-distanced limits. Ditto for my in-law family including Bernie and Barbara Brodsky, Chance Brodsky, Claire McGillem, Cordelia Szaton, Betsy Krugliak, and Seth Krugliak. Our friends Dave Schutter and Jinn Bronwen Lee provided rock-solid counsel, creative support, and unfailing cheer—they are a trifecta of excellence in friendship. Our quarantine pod was an absolute lifesaver. Malcolm MacIver, Kristin Pichaske, Sasha MacIver, and Bianca Gediehn now count as much more than neighbors, more than friends, certainly more than the carpool partners they started out as. There isn't a word powerful enough yet for the kind of deep support we provided each other during an unprecedentedly trying period. While I'm at a loss for words, here's a heartfelt something to my two dearest people ever, my husband, Seth Brodsky, and son, Lev Stewart-Brodsky. I may tell you both constantly how cute and squishy you are, but it's only to belie how extraordinarily lucky I feel to have you both as my bedrocks. I love you.

Joy Goodwin gets her own paragraph! Joy and I formed a writers' group of two for several years, alternately reading and editing each other's work. Joy's input was formative to this project, and I can sense the imprint of her clear, lucid style on every page. She's a superlative writer, reader, and thinker with a lot more good work ahead of her. This book is dedicated to Joy.

NOTES

INTRODUCTION: WHY A BOOK ON SMELL?

1. Daniel Engbar, "Does Poison Gas Smell Good?" *Slate*, August 22, 2006, https://slate.com/news-and-politics/2006/08/what-does-poison-gas-smell-like.html.
2. Tony Phillips, "The Mysterious Smell of Moondust," *NASA Science,* January 30, 2006, https://science.nasa.gov/science-news/science-at-nasa/2006/30jan_smellofmoondust/.
3. David Leonard, "What Does Mars Smell Like?" *Space.com*, June 9, 2016, https://www.space.com/33115-what-does-mars-smell-like.html.
4. Victor Tangermann, "Breaking: Researchers Discover Signs of Life on Venus," *Futurism*, September 14, 2020, https://futurism.com/signs-of-life-venus-phosphine.
5. Alastair Gunn, "What Do the Other Planets Smell Like?" *BBC Science Focus,* https://www.sciencefocus.com/space/what-do-the-other-planets-smell-like/.
6. Diane Ackerman, *A Natural History of the Senses* (New York: Vintage Books, 1990), 54. Lyall Watson, *Jacobson's Organ: And the Remarkable Nature of Smell* (Harmondsworth: Penguin, 1999), 114. Rachel Herz, *The Scent of Desire: Discovering Our Enigmatic Sense of Smell* (New York: Harper Perennial, 2007), 135, 212, 219.
7. Alex Stone, "Smell Turns Up in Unexpected Places," *New York Times*, October 13, 2014, https://www.nytimes.com/2014/10/14/science/smell-turns-up-in-unexpected-places.html.
8. Herz, *The Scent of Desire*, 18, and Jasper H. B. de Groot, et al., "The Knowing Nose: Chemosignals Communicate Human Emotions," *Association for Psychological Science*, November 5, 2012, https://www.psychologicalscience.org/news/releases/the-knowing-nose-chemosignals-communicate-human-emotions.html.

9. "Phantom Smells May Be a Sign of Trouble," *NBC News*, July 10, 2018, https://www.nbcnews.com/health/health-news/phantom-smells-may-be-sign-trouble-n890271.

10. Fiona Macrae, "People who can no longer smell peppermint, fish, rose or leather 'may have only five years left to live,' " *Daily Mail UK*, October 1, 2014, https://www.dailymail.co.uk/health/article-2776651/People-no-longer-smell-peppermint-fish-rose-leather-five-years-left-live.html, and Elizabeth Palermo, "Your Sense of Smell Could Predict When You'll Die," *LiveScience*, October 1, 2014, https://www.livescience.com/48101-loss-of-smell-predicts-mortality.html.

11. Jonathan Reinarz, *Past Scents: Historical Perspectives on Smell* (Urbana, Chicago, and Springfield: University of Illinois Press, 2014), 94–95. See also Miriam Kresh, "Jews and Garlic: Love, Hate, and Confit," Forward, May 1, 2012, https://forward.com/food/155580/jews-and-garlic-love-hate-and-confit/.

12. William Brink and Louis Harris, *The Nego Revolution in America* (New York: Simon and Schuster, 1969), 14.

13. A mutation of the ABCC11 gene that happened 2,000 generations ago has resulted in two differences in bodily smells between ethnic groups. Most East Asians and Koreans lack a chemical in their armpits that produces strong bodily odor; their ear wax is also drier, whiter, and much less smelly than that of Caucasians or people of African descent. Erika Engelhaupt, "What your earwax says about your ancestry," *ScienceNews,* February 14, 2014. https://www.sciencenews.org/blog/gory-details/what-your-earwax-says-about-your-ancestry. See also Bettina Beer, "Geruch und Differenz: Körpergeruch als Kennzeichnen Konstruierter 'Rassischer' Grenzen" ["Smell and Difference: Bodily Odor as an Indicator of Constructed 'Racial' Boundaries]," *Paideuma: Mitteilungen zur Kulturkunde* 46 (2000), 207–30.

14. Abdullah Hamidaddin, "Nose kiss, anyone? How the Gulf Arab greeting has evolved," *Al Arabiya News*, May 4, 2014, http://english.alarabiya.net/en/perspective/features/2014/05/04/Nose-kiss-anyone-How-the-Gulf-Arab-greeting-has-evolved.html.

15. Ackerman, *A Natural History of the Senses*, 23.

16. Kate Fox and Social Issues Research Centre (SIRC), *"The Smell Report: An Overview of Facts and Findings,"* http://www.sirc.org/publik/smell_culture.html.

17. Herz, *The Scent of Desire,* 14–17.

18. Herz, *The Scent of Desire*, 20–21, and Michael Stoddart, *Adam's Nose, and the Making of Humankind* (London: Imperial College Press, 2015), 36–37.

19. Paolo Pelosi, *On the Scent: A Journey Through the Science of Smell*. (Oxford: Oxford University Press, 2016), 59.

20. Herz, *The Scent of Desire*, 234.

21. Herz, *The Scent of Desire*, 210–11.

22. Nicola Twilley, "Will Smell Ever Come to Smartphones?" *New Yorker*, April 27, 2016, https://www.newyorker.com/tech/annals-of-technology/is-digital-smell-doomed.

23. As described on http://smeller.net/.

24. Scott Meslow, " 'Spy Kids 4' Joins Hollywood's Spotty Tradition of Smelly Cinema," *The Atlantic*, August 18, 2011, http://www.theatlantic.com/entertainment/archive/2011/08/spy-kids-4-joins

-hollywoods-spotty-tradition-of-smelly-cinema/243825/, and John Brownlee, "A Brief History of Smell-O-Vision," *Wired*, December 7, 2006, https://www.wired.com/2006/12/a-brief-history-2-2/.

25. Jessica Love, "Why So Few English Words for Odors?" *The American Scholar*, January 16, 2014, https://theamericanscholar.org/why-so-few-english-words-for-odors/#.XXEyoZNKjOR.

26. Farhad Manjoo, "We Have Reached Peak Screen. Now Revolution Is in the Air," *New York Times*, June 17, 2018, https://www.nytimes.com/2018/06/27/technology/peak-screen-revolution.html.

27. Lewis Thomas, *Late Night Thoughts on Listening to Mahler's Ninth Symphony* (New York: Random House Publishing Group, 1984), 42.

How to Read This Book

1. Chris Weller, "10 Different Smells Are Detectable By Your Nose: How Did Popcorn Make The List?," Medical Daily, September 20, 2013, https://www.medicaldaily.com/10-different-smells-are-detectable-your-nose-how-did-popcorn-make-list-257395.

PART ONE: THE NOSE

1. Stoddart, *Adam's Nose,* 25. Piet Vroon with Anton van Amerangen and Hans De Vries, *Smell: The Secret Seducer,* trans. Paul Vincent (New York: Farrar, Straus and Giroux, 1997), 18. Avery Gilbert, *What the Nose Knows: The Science of Scent in Everyday Life* (Fort Collins, CO: Synesthetics, Inc., 2014), 43–44. Pelosi, *On the Scent*, 49–67.

2. Gilbert, *What the Nose Knows*, 78–85.

3. Herz, *The Scent of Desire*, 20–21 and Stoddart, *Adam's Nose,* 36–37. Biochemist and perfume expert Luca Turin argued vociferously against the "shape" theory of smells in favor of his own "vibrational" theory. His research suggests that odorant molecules emit a specific vibration, like sound wavelengths; the olfactory receptor responds to this vibration by firing. Chandler Burr's book *The Emperor of Scent* explores this theory in great and entertaining detail. However the olfactory scientists I consulted for this book agree this theory is largely discredited.

4. Sarah C. P. Williams, "Human Nose Can Detect a Trillion Smells," *AAAS Science*, March 20, 2014, https://www.sciencemag.org/news/2014/03/human-nose-can-detect-trillion-smells, and C. Bushdid, et al, "Humans can discriminate more than 1 trillion olfactory stimuli," *Science* 343, no. 6177 (2014): 1370–72, doi:10.1126/science.1249168. Olfactory experts I consulted emphasized that the trillion-smells figure is easily misinterpreted. As Professor Michael Stoddart explained in emailed correspondence to me: "Theoretically a figure of a trillion is quite possible, but this is based on the number of combinations and permutations of the known nerves in the olfactory system rather than on experimentation. If I asked you how many words you knew, you would likely answer that you could command half a million words in English as an educated person (or a quarter of a million if you were French, or German, or Spanish). You could view this as the maximum. But if you were multilingual you might be able to summon up 1 or 1.5 million words. I think it is not dissimilar with smells. Every one of us can train our sense of smell to recognise tens of thousands of smells;

if you work as a trained perfumier you will have recall of many more than a person who works on a building site, for example. . . . To me, the important thing is the limitless number of smells any one of us could train ourselves to recognise, rather than some upper theoretical maximum."

5. Stoddard, *Adam's Nose,* 32–33.

6. Piet Vroon with Anton van Ameranger and Hans De Vries, *Smell: The Secret Seducer.* Translated by Paul Vincent (New York: Farrar, Straus and Giroux, 1997), 28–29, 31. Stoddard, *Adam's Nose,* 77–79.

7. Weizmann Institute of Science, "An exception to the rule: An intact sense of smell without a crucial olfactory brain structure," *ScienceDaily,* www.sciencedaily.com/releases/2019/11/191111104957.htm, accessed December 30, 2020.

8. Jennifer Pluznick, "You smell with your body, not just your nose," filmed November 2016 at TED-MED, Palm Springs, CA, https://www.ted.com/talks/jennifer_pluznick_you_smell_with_your_body_not_just_your_nose#t-412793. Alex Stone, "Smell Turns Up in Unexpected Places," *New York Times*, October 13, 2014, https://www.nytimes.com/2014/10/14/science/smell-turns-up-in-unexpected-places.html. Veronique Greenwood, "How Bacteria Help Regulate Blood Pressure," *Quanta*, November 30, 2017, https://www.quantamagazine.org/how-bacteria-help-regulate-blood-pressure-20171130/. Patrick Caughill, "Our Sense of Smell Provides a New Way to Battle Spinal Cord Injuries," December 11, 2017, https://futurism.com/neoscope/sense-smell-provides-new-way-battle-spinal-cord-injuries.

9. Robert Muchembled, *Smells: A Cultural History of Odours in Early Modern Times*, trans. Susan Pickford (Cambridge, UK, and Medford, Massachusetts: Polity, 2020), 8, and Herz, *The Scent of Desire,* 33.

10. Vroon, *Smell*, 21–22, and Herz, *The Scent of Desire,* 33–35.

11. Herz, *The Scent of Desire,* 33–34.

12. Christopher Bergland, "How Do Nostalgic Scents Get Woven Into Long-Term Memories?" *Psychology Today*, December 25, 2017, https://www.psychologytoday.com/us/blog/the-athletes-way/201712/how-do-nostalgic-scents-get-woven-long-term-memories.

13. Johan Willander and Maria Larsson, "Smell your way back to childhood: autobiographical odor memory," *Psychonomic Bulletin & Review* 13, no. 2 (2006): 240–44, doi:10.3758/bf03193837, https://pubmed.ncbi.nlm.nih.gov/16892988/.

14. Philip Perry, "Think you have only 5 senses? You've actually got about 14 to 20," *BigThink*, May 2, 2018, https://bigthink.com/philip-perry/think-you-have-only-5-senses-its-actually-a-lot-more-than-that.

15. Constance Classen, *Worlds of Sense: Exploring the Senses in History and Across Cultures* (London and New York: Routledge, 1993), 5.

16. Lynne Peeples. "Making Scents of Sounds," *Scientific American* 302 (April 2020): 28–29.

17. Pelosi, *On the Scent*, 14–15, and Gilbert, *What the Nose Knows*, 92–93.

18. Herz, *The Scent of Desire,* 47.

19. Classen, *Worlds of Sense*, 2.

20. Reinarz, *Past Scents,* 8–15, and Classen, *Worlds of Sense*, 2–7.

21. 20 percent and 80 percent figures: Reinarz, *Past Scents*, 80. Market size of flavors and fragrance market is expected to reach $36.6 billion by 2024 according to Flavor and Fragrance Market Report: Trends, Forecast and Competitive Analysis (May 2019): https://www.researchandmarkets.com/reports/4790918/flavor-and-fragrance-market-report-trends?utm_source=CI&utm_medium=PressRelease&utm_code=tq4w85&utm_campaign=1271847+-+Global+Flavor+%26+Fragrance+Market+to+Reach+%2436.6+Billion+by+2024&utm_exec=joca220prd.

22. Classen, *Worlds of Sense*, 59.

23. Mark Bradley, ed., *Smell and the Ancient Senses* (London and New York: Routledge, 2015), 3.

24. Robert G. Walker, "A Sign of the Satirist's Wit: The Nose in Tristram Shandy," *Ball State University Forum* 19, no. 2 (January 1978): 52–54.

25. Eddy Portnoy, "The Nose Knows: George Jabet's *Nasology*," interview in "Beach Bodies: A History of American Physique," *BackStory,* July 13, 2015, audio, 1:58: https://americanarchive.org/catalog/cpb-aacip_532-kk94748554, and show notes: https://www.backstoryradio.org/blog/the-nose-knows.

26. Anthony Kingston, "The Fundamentals of Chinese Face-Reading," *Professional Beauty*, no. Mar/Apr 2017 (April 1, 2017): 66–68.

27. Annick Le Guérer, *Scent: The Essential and Mysterious Powers of Smell*, trans. Richard Miller (New York: Kodansha America, 1994), 11.

28. "Justinian II: Byzantine Emperor," Encyclopedia Britannica, last updated November 27, 2020, https://www.britannica.com/biography/Justinian-II.

29. G. Sperati, "Amputation of the Nose throughout History," *Acta Otorhinolaryngologica Italica (Testo Stampato)* 29, no. 1 (Jan. 2009): 44–50, https://www.ncbi.nlm.nih.gov/pmc/articles/PMC2689568/.

30. Cathy Newman, with photography by Robb Kendrick, *Perfume: The Art and Science of Scent* (Washington, DC: National Geographic Society, 1998), 85–95, 100. Gilbert, *What the Nose Knows*, 26–28, 37.

31. Rinni Bhansali, "Stanford grads' startup Aromyx aims to quantify taste and smell," *The Stanford Daily*, October 11, 2019, https://www.stanforddaily.com/2019/10/11/stanford-grads-startup-aromyx-aims-to-quantify-taste-and-smell/.

32. Charlotte Ouwerkerk, "Dutch robots help make cheese, 'smell' the roses," *The Jakarta Post* (via Agence-France Presse), January 26, 2018, https://www.thejakartapost.com/life/2018/01/25/dutch-robots-help-make-cheese-smell-the-roses.html.

33. Greg Nichols, "Robots will soon be able to taste and smell your bad cooking," ZDNet, December 15, 2017, https://www.zdnet.com/article/robots-will-soon-be-able-to-taste-and-smell-your-bad-cooking/, and American Chemical Society, "Bioelectronic 'nose' can detect food spoilage by sensing the smell of death," *ScienceDaily*, www.sciencedaily.com/releases/2017/12/171206122457.htm, accessed December 30, 2020.

34. Nanoscent Labs is piloting breathalyzer-style tests to detect COVID-19 within minutes: https://youtu.be/eoL_DQh9L98. This is effectively an electronic version of COVID-sniffing dogs in airports. Elian Peltier, "The Nose Needed for This Coronavirus Test Isn't Yours. It's a Dog's," *New York Times*, September 23, 2020, https://www.nytimes.com/2020/09/23/world/europe/finland-dogs-airport-coronavirus.html.

35. Vanessa Zainzinger, "Rosa Biotech gets biosensors on the nose," *Chemistry World*, September 14, 2020, https://www.chemistryworld.com/news/rosa-biotech-gets-biosensors-on-the-nose/4012417.article.

36. Ed Yong, "Scientists Stink at Reverse-Engineering Smells," *The Atlantic*, November 14, 2016, https://www.theatlantic.com/science/archive/2016/11/how-to-reverse-engineer-smells/507608/.

37. Salk Institute, "What's that smell? Scientists find a new way to understand odors: A mathematical model reveals a map for odors from the natural environment," *ScienceDaily*, August 29, 2018, www.sciencedaily.com/releases/2018/08/180829143816.htm, accessed December 29, 2020.

PART TWO: THE SMELLS

1. Flowery and Herbal

Petrichor

1. Cynthia Barnett, "Making Perfume From the Rain," *The Atlantic*, April 22, 2015, https://www.theatlantic.com/international/archive/2015/04/making-perfume-from-the-rain/391011/.

2. American Chemical Society, "Video: Petrichor, the smell of rain," Phys.org, April 3, 2018, video: https://phys.org/news/2018-04-video-petrichor.html.

3. Jennifer Chu, "Rainfall can release aerosols, study finds," *MIT News*, January 14, 2015, https://news.mit.edu/2015/rainfall-can-release-aerosols-0114.

4. Daisy Yuhas, "Storm Scents: It's True, You Can Smell Oncoming Summer Rain," July 18, 2012, https://www.scientificamerican.com/article/storm-scents-smell-rain/.

5. Simon Cotton, "Geosmin: The Smell of the Countryside," *Molecule of the Month*, August 2009, http://www.chm.bris.ac.uk/motm/geosmin/geosminh.htm, and Paul Simons, "Camels act on a hump," *The Guardian*, March 5, 2003, https://www.theguardian.com/science/2003/mar/06/science.research.

Roses

1. Classen, *Worlds of Sense*, 17–18 and Jennifer Potter, *The Rose: A True History* (London: Atlantic Books, 2010), 50. Potter notes that, while this is the story most commonly told, the original source describes a deluge of "violets and other flowers", not roses.

2. Frederick E. Brenk, *Clothed in Purple Light: Studies in Vergil and in Latin Literature* (Stuttgart: Franz Steiner, 1999), 88.

3. Potter, *The Rose*, 244.

4. Potter, *The Rose,* 260.

5. Peter Bernhardt, *The Rose's Kiss: A Natural History of Flowers* (Washington, DC, and Covelo, CA: Island Press and Shearwater Books, 1999), 120.

6. Potter, *The Rose,* 248–250. Potter is quoting the author Vita Sackville-West who visited Persia in the nineteenth century and relates this anecdote.

7. Ackerman, *A Natural History of the Senses*, 34.

8. Ackerman, *A Natural History of the Senses*, 60.

9. Classen, *Worlds of Sense*, 24.

10. Mandy Aftel, *Essence & Alchemy: A Book of Perfume* (Layton, UT: Gibbs Smith, 2008), 110.

11. Steve Connor, "Scientists discover why some roses smell sweeter than others—and how to improve the scent," *Independent UK*, July 2, 2015, https://www.independent.co.uk/news/science/scientists-discover-why-some-roses-smell-sweeter-than-others-and-how-to-improve-the-scent-10362519.html. Note that it's roses bred for modern, cut-flower markets that are most likely to be odorless. Certain rose breeders like David Austen have long emphasized breeding roses for their fragrance and have done so successfully before the discovery of this enzyme.

Jasmine

1. Mandy Aftel, *Fragrant: The Secret Life of Scent* (New York: Riverhead Books, 2014), 205–207.

2. Nigel Groom, *The Perfume Handbook* (London: Chapman & Hall, 1992), 97.

3. Groom, *The Perfume Handbook*, 69.

4. Newman, *Perfume*, 67.

5. Newman, *Perfume*, 76–77. Aftel, *Fragrant,* 208–209. Marie-Christine Grasse, *Jasmine: Flower of Grasse* (Parkstone Publishers & Musée Internationale de la Perfume, 1996), 46–50.

6. Newman, *Perfume,* 72.

7. Aftel, *Essence & Alchemy,* 111.

8. Grasse, *Jasmine,* 85–87.

9. Newman, *Perfume,* 35.

Fresh-Cut Grass

1. Melissa Breyer, "What Fresh Cut Grass Is Saying With Its Scent," *Treehugger*, May 17, 2019, https://www.treehugger.com/what-fresh-cut-grass-saying-its-scent-4855746.

2. Melissa Breyer, "How Trees Talk to Each Other and Share Gifts," *Treehugger*, October 11, 2018, https://www.treehugger.com/how-trees-talk-each-other-and-share-gifts-4856994.

Line-Dried Laundry

1. Cara Giaimo, "How Line-Dried Laundry Gets That Fresh Smell," *New York Times*, May 29, 2020, https://www.nytimes.com/2020/05/29/science/laundry-smell-line.html.

2. Katharine Wroth, "A surprising sneak peek at the clothesline revolution," *Grist*, November 13, 2009, https://grist.org/article/2009-11-12-alex-lee-clothesline-revolution/.

Vanilla

1. Nigel Groom, *The Perfume Handbook* (London: Chapman & Hall, 1992), 244–45. Patricia Rain, *Vanilla: The Cultural History of the World's Favorite Flavor and Fragrance* (New York: Jeremy P. Tarcher, 2004), 5.

2. Rain, *Vanilla*, 5. Tim Ecott, *Vanilla: Travels in Search of the Ice Cream Orchid* (New York: Grove Press, 2004), 20. During fact-checking Patricia Rain informed me via email of evidence suggesting an alternate origin for Tahitian vanilla: in her words "a cross between Vanilla planifolia and Vanilla odorata, one of the many sylvestre vanilla varieties in Mexico. It was likely already hybridized before leaving Mexico for the Philippines. After it arrived in Tahiti, tahitensis was further manipulated."

3. Rain, *Vanilla*, 19–37. Ecott, *Vanilla*, 6–10.

4. Ecott, *Vanilla*, 22–23.

5. Rain, *Vanilla*, 7–10. Ecott, *Vanilla*, 62–64.

6. Rain, *Vanilla*, 9.

7. Robert Krulwich, "The Little Boy Who Should've Vanished, but Didn't," *National Geographic*, June 16, 2015, https://www.nationalgeographic.com/science/phenomena/2015/06/16/the-little-boy-who-shouldve-vanished-but-didnt/. The original source of this research is Ecott, *Vanilla*, 123-135.

8. Ecott, *Vanilla*, 122–25.

9. Ecott, *Vanilla*, 131–48.

10. Anna Diamond, "Make Thomas Jefferson's Recipe for Ice Cream," *Smithsonian Magazine*, July/August 2020, https://www.smithsonianmag.com/history/thomas-jefferson-ice-cream-recipe-180975200/. See also Rain, *Vanilla*, 63—69.

11. Amanda Fortini, "The White Stuff: How Vanilla Became Shorthand for Bland," *Slate*, August 10, 2005, https://slate.com/human-interest/2005/08/how-vanilla-became-shorthand-for-bland.html.

12. Rain, *Vanilla*, 132–56.

13. Herz, *The Scent of Desire,* 40.

14. "Scent of vanilla helps to ease pain: Japanese researchers," *Kyodo News*, July 13, 2019, https://english.kyodonews.net/news/2019/07/50384c5dcce4-scent-of-vanilla-helps-to-ease-pain-japanese-researchers.html.

15. W. H. Redd et al., "Fragrance administration to reduce anxiety during MR imaging," *Journal of Magnetic Resonance Imaging* 4, no. 4 (1994): 623–26, doi:10.1002/jmri.1880040419, https://pubmed.ncbi.nlm.nih.gov/7949692/.

16. Karen Eisenbraun, "Vanilla Aromatherapy Benefits," *OurEverydayLife*, August 14, 2017, https://oureverydaylife.com/149191-vanilla-aromatherapy-benefits.html.

17. Kat Chow, "When Vanilla Was Brown And How We Came To See It As White," *Code Switch* on

NPR, March 2, 2014, https://www.npr.org/sections/codeswitch/2014/03/23/291525991/when
-vanilla-was-brown-and-how-we-came-to-see-it-as-white.

Sweet Woodruff

1. Peter Wagner, "Deutschland ist Waldmeister!" *Der Spiegel*, May 9, 2010, https://www.spiegel.de
   /consent-a-?targetUrl=https%3A%2F%2Fwww.spiegel.de%2Fkultur%2Fgesellschaft%2
   Fmaibowle-deutschland-ist-waldmeister-a-693215.html (in German).
2. "Maibowle," ChefKoch, January 15, 2007, https://www.chefkoch.de/rezepte/668471168850775
   /Maibowle.html (in German).

Bitter Almonds

1. Diana Lutz, "Beware the smell of bitter almonds," The Source (Washington University in St.
   Louis), July 20, 2010, https://source.wustl.edu/2010/07/beware-the-smell-of-bitter-almonds/.
2. Lutz, "Beware the smell of bitter almonds."
3. Peggy Trowbridge Filippone, "What are Bitter or Lethal Almonds?" *The Spruce Eats*, October 13,
   2019, https://www.thespruceeats.com/what-are-bitter-almonds-1806996.
4. Daniel Engbar, "Does Poison Gas Smell Good?"
5. Sarah Mohr, "Datura," University of Wisconsin-Madison Master Gardener's Program, February
   11, 2019, https://mastergardener.extension.wisc.edu/article/datura/#:~:text=The%20flowers
   %20exude%20a%20pleasant,honey%20bees%20and%20other%20insects.
6. John Rogers, "VEGGIE FRIGHT: Deadly plant hemlock which can kill a human with ONE bite
   found on beach in the UK," *The Sun UK*, March 10, 2018, https://www.thesun.co.uk/news/5774916
   /hemlock-uk-deadly-plant-dogs-cornwall-storm-emma/, and Eliza Strickland, "Plant That Produced Ritual Death-Smiles May've Given Homer a Neat Phrase," *Discover Magazine*, June 3, 2009,
   https://www.discovermagazine.com/health/plant-that-produced-ritual-death-smiles-mayve
   -given-homer-a-neat-phrase.
7. "Belladonna," Encyclopedia.com, updated December 22, 2020, https://www.encyclopedia.com
   /medicine/drugs/pharmacology/belladonna.

Cinnamon

1. Jack Turner, *Spice: The History of a Temptation* (New York: Knopf, 2004), 148.
2. Aftel, *Fragrant,* 39.
3. Aftel, *Fragrant,* 35º36.
4. Turner, *Spice,* 232.
5. "Cinnamaldehyde," Encyclopedia.com, updated December 22, 2020, https://www.encyclopedia
   .com/science/academic-and-educational-journals/cinnamaldehyde.
6. Turner, *Spice,* 206.

7. Turner, *Spice,* 149–50.

8. Stephanie Butler, "The Medieval History of the Christmas Cookie," History.com, December 18, 2013 (updated August 31, 2018), https://www.history.com/news/the-medieval-history-of-the-christmas-cookie.

9. Peggy Trowbridge Filippone, "What is Cassia? All About Cinnamon's Cousin," *The Spruce Eats*, updated February 8, 2019, https://www.thespruceeats.com/what-is-cassia-1807003.

10. Turner, *Spice,* 203, 208–209, 217.

11. Turner, *Spice*, 185–86.

12. Turner, *Spice*, 222.

13. Turner, *Spice*, 188.

Hot Chocolate

1. American Chemical Society, "The smell of dark chocolate, demystified," *AAAS EurekaAlert!*, May 8, 2019, https://www.eurekalert.org/pub_releases/2019-05/acs-tso050319.php.

2. Carrie Arnold, "The Sweet Smell of Chocolate: Sweat, Cabbage and Beef," *Scientific American,* October 31, 2011, https://www.scientificamerican.com/article/sensomics-chocolate-smell/.

3. Sarah Moss and Alexander Badenoch, *Chocolate: A Global History* (London: Reaktion Books, 2009), 12.

4. Moss and Badenoch, *Chocolate*, 13.

5. Moss and Badenoch, *Chocolate*, 18.

6. Moss and Badenoch, *Chocolate*, 22.

7. Moss and Badenoch, *Chocolate*, 24–25.

8. Moss and Badenoch, *Chocolate*, 31.

9. Moss and Badenoch, *Chocolate*, 33.

10. Moss and Badenoch, *Chocolate*, 40.

11. Moss and Badenoch, *Chocolate*, 57, 62–63.

12. "Child Labor and Slavery in the Chocolate Industry," Food Empowerment Project, https://foodispower.org/human-labor-slavery/slavery-chocolate/.

13. Melissa Clark, "Everything You Don't Know About Chocolate," *The New York Times*, February 11, 2020, https://www.nytimes.com/2020/02/11/dining/chocolate-bar.html?referringSource=articleShare.

3. Savory

Bacon

1. Gareth May, "Why do we love the smell of bacon so?" *The Telegraph* UK, March 16, 2015, https://www.telegraph.co.uk/foodanddrink/11470099/Why-do-we-love-the-smell-of-bacon-so.html.

2. "Wake N Bacon: Bacon Cooking Alarm Clock," Shark Tank Products, June 6, 2011, https://allsharktankproducts.com/shark-tank-products-food-and-drink/wake-n-bacon-bacon-cooking-alarm-clock/.

3. Russ Parsons, "Why does bacon smell so good? It's all in the chemistry," *Los Angeles Times*, April 17, 2014, https://www.latimes.com/food/dailydish/la-dd-calcook-why-does-bacon-smell-so-good-its-all-in-the-chemistry-20140417-story.html.

4. Maria Godoy, "Does Bacon Really Make Everything Better? Here's The Math," *The Salt* on NPR, October 25, 2013, https://www.npr.org/sections/thesalt/2013/10/25/240556687/does-bacon-really-make-everything-better-here-s-the-math.

Durian

1. Jia-Xiao Li, Peter Schieberle, and Martin Steinhaus, "Characterization of the Major Odor-Active Compounds in Thai Durian (*Durio zibethinus* L. 'Monthong') by Aroma Extract Dilution Analysis and Headspace Gas Chromatography–Olfactometry," *Journal of Agricultural and Food Chemistry* 60, no. 45 (2012): 11253–62, doi: 10.1021/jf303881k, https://pubs.acs.org/doi/abs/10.1021/jf303881k?prevSearch=durian&searchHistoryKey=#.

2. Joseph Stromberg, "Why Does the Durian Fruit Smell So Terrible?," *Smithsonian Magazine,* November 30, 2012, https://www.smithsonianmag.com/science-nature/why-does-the-durian-fruit-smell-so-terrible-149205532/.

3. Randall Munroe, "What's the world's worst smell?" *Chicago Tribune* (via *The New York Times*), February 17, 2020, https://www.chicagotribune.com/featured/sns-nyt-worst-smell-in-the-world-20200217-xywylvisqfddnpx2cbbkvw2wwa-story.html.

4. Jia-Xiao Li, Peter Schieberle, and Martin Steinhaus, "Insights into the Key Compounds of Durian (*Durio zibethinus* L. 'Monthong') Pulp Odor by Odorant Quantitation and Aroma Simulation Experiments," *Journal of Agricultural and Food Chemistry* 65, no. 3 (2017): 639–47, doi:10.1021/acs.jafc.6b05299, https://pubs.acs.org/doi/full/10.1021/acs.jafc.6b05299?hootPostID=3ac35a02cbe0d732673c81d34f813bac.

5. Munroe, "What's the world's worst smell?"

6. Alfred Russel Wallace, *The Malay Archipelago: The Land of the Orang-utan and the Bird of Paradise; a Narrative of Travel, with Studies of Man and Nature* (London and New York: Macmillan and Co., 1894), 57.

Stinky Cheese

1. "World's 'Smelliest' Cheese Named," *The Telegraph* UK, November 25, 2004, https://www.telegraph.co.uk/news/1477473/Worlds-smelliest-cheese-named.html.

2. Richard Sutton, "What Makes Stinky Cheese Stinky?" *Culture Magazine*, January 19, 2015, https://culturecheesemag.com/ask-the-monger/makes-stinky-cheese-stinky/.

3. Andrew Dalby, *Cheese: A Global History* (London: Reaktion Books, 2009), 89.

4. Bronwen Percival and Francis Percival, *Reinventing the Wheel: Milk, Microbes and the Fight for Real Cheese* (Oakland: University of California Press, 2017), 30–41. Also Dalby, *Cheese*, 33.

5. Dalby, *Cheese*, 37.

6. Dalby, *Cheese*, 82–83.

7. Dalby, *Cheese*, 15–16.

8. Matt Colangelo, "A Desperate Search for Casu Marzu, Sardinia's Illegal Maggot Cheese," *Food & Wine*, updated October 14, 2015, https://www.foodandwine.com/news/desperate-search-casu-marzu-sardinias-illegal-maggot-cheese. Marc Frauenfelder, *The World's Worst: A Guide to the Most Disgusting, Hideous, Inept, and Dangerous People, Places, and Things on Earth* (San Francisco: Chronicle Books, 2005), 22, https://books.google.com/books?id=iUQ5GNNOFAEC&printsec=frontcover&dq=isbn:9780811846066&hl=en&newbks=1&newbks_redir=0&sa=X&ved=2ahUKEwivv8qky-rnAhUEa60KHaPbDW4Q6AEwAHoECAAQAg#v=onepage&q=casu&f=false. Dalby, *Cheese*, 85.

9. Brian Handwerk, "What Stinky Cheese Tells Us About the Science of Disgust," *Smithsonian Magazine,* October 3, 2017, https://www.smithsonianmag.com/science-nature/what-stinky-cheese-tells-us-about-disgust-180965017/.

10. "Food—Delicious Science: Backward Smelling," PBS, posted online May 15, 2017, video: https://www.youtube.com/watch?v=ylmdlaSHQ2I&feature=emb_logo&ab_channel=PBS.

Asafoetida

1. "Asafetida (*Ferula assa-foetida* L.)," Gernot Katzer's Spice Pages, http://gernot-katzers-spice-pages.com/engl/Feru_ass.html.

2. Chip Rossetti, " 'Devil's Dung': The World's Smelliest Spice," *Aramco World*, July/August 2009, https://archive.aramcoworld.com/issue/200904/devil.s.dung-the.world.s.smelliest.spice.htm.

Tobacco

1. Andrew Wike, "Know Your Pipe Tobacco: Blending Components," SmokingPipes.com, December 23, 2016, https://www.smokingpipes.com/smokingpipesblog/single.cfm/post/know-your-pipe-tobacco-blending-components.

2. "Black Frigate" product page, https://www.smokingpipes.com/pipe-tobacco/cornell-diehl/Black-Frigate-2oz/product_id/225.

3. "War Horse" product page, https://www.smokingpipes.com/pipe-tobacco/war-horse/Bar-1.75oz/product_id/220589.

4. "Sun Bear" product page, https://www.smokingpipes.com/pipe-tobacco/cornell-diehl/Sun-Bear-2oz/product_id/336163. For more on casing versus aromatic tobaccos, see Chuck Stanion, "A Closer Look At Aromatic Pipe Tobacco," *SmokingPipes.com*, August 7, 2020, https://www.smokingpipes.com/smokingpipesblog/single.cfm/post/closer-look-aromatic-pipe-tobacco.

5. "How to Pack & Light A Pipe," SmokingPipes.com, https://www.smokingpipes.com/information/howto/packing.cfm.

6. Iain Gately, *Tobacco: A Cultural History of How an Exotic Plant Seduced Civilization* (New York: Grove Press, 2001), 2–3, 19, 23, 28. After the Mayans, the imperialist Aztecs brought tobacco to every people they conquered, disseminating the weed to the northernmost reaches of the American continent by 2500 BCE—see Gately, *Tobacco,* 13.

4. Earthy

Truffles

1. Ryan Jacobs, *The Truffle Underground: A Tale of Mystery, Mayhem and Manipulation in the Shadowy Market of the World's Most Expensive Fungus* (New York: Clarkson Potter, 2019), 69.
2. Jacobs, *The Truffle Underground*, 71–77.
3. Jacobs, *The Truffle Underground*, 51–58.
4. Jacobs, *The Truffle Underground*, 89.
5. Jacobs, *The Truffle Underground*, 236.

Wine

Unless otherwise noted, all points in this chapter derive from my interview with sommelier Alex Ring of Oriole in Chicago.

1. Bianca Bosker, *Cork Dork: A Wine-Fueled Adventure Among the Obsessive Sommeliers, Big Bottle Hunters and Rogue Scientists Who Taught Me to Live for Taste* (New York: Penguin Books, 2017), 46–49.
2. From conversations with Howard Spiegelman, wine instructor.
3. Bosker, *Cork Dork*, 186–87. See also "Wine for the Confused" hosted by John Cleese, The Food Network, video: https://www.youtube.com/watch?v=sHnz6KoYw_A&ab_channel=RockBandito.
4. Madeline Puckett, "Brunello di Montalcino: Well Worth The Wait," *Wine Folly*, August 12, 2016 (updated March 27, 2020), https://winefolly.com/deep-dive/brunello-di-montalcino-its-worth-the-wait/.

Cannon Fire

1. Benjamin Sobieck, "What's that Smell? Cordite vs. Gunpowder vs. Propellant," *The Writer's Guide to Weapons: A Practical Reference for Using Firearms and Knives in Fiction* (Cincinnati, OH: F+W Media, 2014), https://crimefictionbook.com/2015/04/30/whats-the-smell-cordite-vs-gunpowder-vs-propellant/.
2. Smith, *The Smell of Battle*, 48–49, 73–75, 78–79. The modern-day Institute for Creative Tech

(ICT) uses virtual reality smells of war to train soldiers and inoculate them against smells' emotionally triggering effects—see Herz, *The Scent of Desire,* 232.

3. Vroon, *Smell*, 6–7, 10. Miasma theory dates as far back as Seneca according to Annick Le Guérer, *Scent: The Essential and Mysterious Powers of Smell,* Translated by Richard Miller (New York: Kodansha America, 1994), 41.

4. Vroon, *Smell*, 9. See also Le Guérer, *Scent*, 66, 74–75.

5. Le Guérer, *Scent*, 66–74. See also Melanie A. Kiechle, *Smell Detectives: An Olfactory History of Nineteenth-Century Urban America* (Seattle: University of Washington Press, 2017), 54, and Classen, *Worlds of Sense*, 21.

6. Alain Corbin, *The Foul and the Fragrant: Odor and the French Social Imagination* (Cambridge: Harvard University Press, 1986), 64.

7. Le Guérer, *Scent*, 80–83.

8. Le Guérer, *Scent*, 78.

9. Rodolphe el-Khoury, "Polish and Deodorize: Paving the City in Late Eighteenth-Century France," in *The Smell Culture Reader*, ed. Jim Drobnick (New York: Berg, 2006), 18–27.

Melting Permafrost

1. "What does thawed permafrost smell like?" PolarTREC, June 14, 2019, https://www.youtube.com/watch?v=a3jv6OaxM24&ab_channel=PolarTREC.

2. Brian Resnick, "Scientists feared unstoppable emissions from melting permafrost. They may have already started," *Vox*, December 12, 2019, https://www.vox.com/energy-and-environment/2019/12/12/21011445/permafrost-melting-arctic-report-card-noaa.

3. University of Copenhagen, "A new permafrost gas mysterium," Phys.org, August 27, 2018, https://phys.org/news/2018-08-permafrost-gas-mysterium.html.

Tea

1. Helen Saberi, *Tea: A Global History* (London: Reaktion Books, 2010), 8. In an email exchange, Saberi notes that "in Poland they call the tea something entirely different—*herbata* (from the Latin *herba thea*, meaning "tea herb.")

2. Saberi, *Tea*, 10.

3. Saberi, *Tea*, 13, 15, 18.

4. Saberi, *Tea*, 27.

5. Saberi, *Tea*, 29.

6. Saberi, *Tea*, 42, 46.

7. Saberi, *Tea*, 58–60.

8. Saberi, *Tea*, 66–67, 71.

9. Saberi, *Tea*, 76–77.

10. Saberi, *Tea*, 81.
11. Saberi, *Tea*, 87, 93.
12. Saberi, *Tea*, 102–103.
13. Saberi, *Tea*, 123.

5. Resinous

Freshly Sharpened Pencils

1. Walter Benjamin, *Berliner Chronik* (Berlin: Karl-Maria Guth, 2016), 61. Translation is mine.
2. Henry Petroski, *The Pencil: A History of Design and Circumstance* (London and Boston: Faber and Faber, 2003), 201–206.
3. Petroski, *The Pencil*, 34, 47, 61. Caroline Weaver, *The Pencil Perfect: The Untold Story of a Cultural Icon* (Berlin: Gestalten, 2017), 7–10, 13.
4. Weaver, *The Pencil Perfect*, 22.
5. Petroski, *The Pencil*, 117.
6. Heather Schwedel, "Why Do Erasers Suck at Erasing?" *The Atlantic,* October 2, 2014, https://www.theatlantic.com/technology/archive/2014/10/why-do-erasers-suck-at-erasing/381025/.

Oud

1. "Ranjatai—The Most Famous piece of Aloeswood," KyaraZen.com, February 8, 2013, https://www.kyarazen.com/ranjatai-the-most-famous-piece-of-aloeswood/, and Kiyoko Morita, *The Book of Incense: Enjoying the Traditional Art of Japanese Scents* (Tokyo and New York: Kodansha International, 1992), 34.
2. Ali Mohamed Al-Woozain, "Scent From Heaven: On the Trail of Oud," Al Jazeera, March 19, 2016, documentary film: https://interactive.aljazeera.com/aje/2016/oud-agarwood-scent-from-heaven/scent-from-heaven-watch.html.
3. Armina Ligaya, "For a Coveted Resin, the Scent of Rarity Takes Hold," *The New York Times,* April 28, 2011, https://www.nytimes.com/2011/04/28/world/middleeast/28iht-M28C-PERFUME.html.
4. Kiyoko Morita, *The Book of Incense: Enjoying the Traditional Art of Japanese Scents* (Tokyo and New York: Kodansha International, 1992), 40. See also Aileen Gatten, "A Wisp of Smoke: Scent and Character in *The Tale of Genji,*" in *The Smell Culture Reader,* ed. Jim Drobnick (New York: Berg, 2006), 331.
5. Morita, *The Book of Incense*, 42–43.
6. Françoise Aubaile-Sallenave, "Bodies, Odors & Perfumes in Arab-Muslim Societies," in *The Smell Culture Reader,* ed. Jim Drobnick (New York: Berg, 2006), 391.

Camphor

1. Reinarz, *Past Scents,* 62–63. "Camphor," *Encyclopedia Britannica*, https://www.britannica.com/science/camphor. R. A. Donkin, *Dragon's Brain Perfume: An Historical Geography of Camphor* (Leiden and Boston: Brill, 1999). On page 178, Donkin also describes how some camphor hunters kitted themselves up in war regalia, as if battling the camphor trees to give up their prize.
2. Sarah Wisseman, "Preserved for the afterlife," *Nature* 413 (October 23, 2001): 783–84. https://doi.org/10.1038/35101673, and Donkin, *Dragon's Brain Perfume*, 85.
3. Donkin, *Dragon's Brain Perfume,* 99.
4. Robert Kennedy Duncan, *Some Chemical Problems of Today* (New York and London: Harper & Brothers Publishers, 1911), 128.
5. Francis Galton, "Arithmetic by Smell," *Psychological Review* 1 (1894), http://www.galton.org/essays/1890-1899/galton-1894-smell.pdf.
6. Donkin, *Dragon's Brain Perfume,* 102–103.

Frankincense

1. Clint Pumphrey, "What are Frankincense and Myrrh?" *How Stuff Works,* updated October 16, 2018, https://science.howstuffworks.com/life/botany/question283.htm.
2. Aftel, *Fragrant,* 138.
3. Aftel, *Fragrant,* 133. Classen, *Worlds of Sense*, 52.
4. "Frankincense—much more than a fragrant smell," University of Birmingham School of Chemistry, December 21, 2017, https://www.birmingham.ac.uk/schools/chemistry/news/2017/frankincense-much-more-than-fragrant-smell.aspx. Carmen Drahl, "What are frankincense and myrrh and why is their smell so mystical?" *Chemical & Engineering News* 86, no. 51 (December 22, 2008), https://cen.acs.org/articles/86/i51/Frankincense-Myrrh.html.
5. "St. Blaise," *Encyclopedia Britannica*, https://www.britannica.com/biography/Saint-Blaise.
6. Alfred Eldersheim, "Why Gold, Frankincense and Myrrh?" *Christianity.com*, May 26, 2010, https://www.christianity.com/jesus/birth-of-jesus/star-and-magi/why-gold-frankincense-and-myrrh.html.

Myrrh

1. Marsha McCulloch, MS, RD, "11 Surprising Benefits and Uses for Myrrh Oil," Healthline, January 4, 2019, https://www.healthline.com/nutrition/myrrh-oil#TOC_TITLE_HDR_5.
2. Kathleen Martin, *The Book of Symbols,* ed. Ami Ronnberg (Germany: Taschen, 2010), 726.
3. Song of Songs 5:4-5, New International Version of the Bible, https://biblehub.com/songs/5-5.htm.
4. Nigel Groom, *Frankincense and Myrrh: A Study of the Arabian Spice Trade* (Harlow, Essex, UK, and Beirut: Longman Group and Librairie du Liban, 1981), 122.

5. Groom, *Frankincense and Myrrh*, 20.

6. Reinarz, *Past Scents,* 55–56.

6. Funky

Skin

1. F. Bryant Furlow, "The Smell of Love," *Psychology Today*, March 1, 1996 (updated June 9, 2016), https://www.psychologytoday.com/us/articles/199603/the-smell-love.

2. Herz, *The Scent of Desire,* 164.

3. Herz, *The Scent of Desire*, 126, 128.

4. Herz, *The Scent of Desire*, 119.

5. Herz, *The Scent of Desire*, 129, 139.

6. Herz, *The Scent of Desire*, 132.

7. Stoddart, *Adam's Nose*, 52–53.

8. "Alan Young on how 'Mister Ed' really talked," EmmyTVLegends.org, video, https://www .youtube.com/watch?v=STTdvkwppBY&ab_channel=FoundationINTERVIEWS. According to this interview with actor Alan Young, producers got the actual Mister Ed to talk by sticking a nylon thread under his lip. They'd yank it on cue, and the horse would try to dislodge it, flaring his lips in a talky-looking way.

9. Stoddart, *Adam's Nose*, 19–20 and 84–90. Stoddart refutes the theory of a still-functioning human VNO on pages 110–114. He also describes a crucial mutation in early hominids, nicknamed ADAM, that severed the VNO's connection to the hominid brain. Stoddart calls this mutation "a defining moment in human olfactory evolution" and names his book after it. In personal correspondence, Prof. Stoddart referred me to Richard Doty, *The Great Pheromone Myth* (Baltimore: Johns Hopkins University Press, 2010), which in Stoddart's words "discusses in forensic detail the purported evidence for and against human pheromones and concludes unequivocally that there is no evidence that humans communicate by pheromones."

10. Herz, *The Scent of Desire,* 214.

11. Rachael Rettner, "Body's Response to Disease Has a Smell, Study Suggest," *LiveScience,* January 24, 2014, https://www.livescience.com/42836-smell-sickness-immune-system.html.

12. Herz, *The Scent of Desire,* 135, 212, 219. See also Ackerman, *A Natural History of the Senses,* 54. Susan Scutti, "Experimental technology can 'smell' disease on your breath," CNN Health, November 7, 2017, https://edition.cnn.com/2017/11/07/health/na-nose-disease-smell-technology /index.html. Bukola Adebayo, "Dogs can sniff out malaria parasites on your clothes," CNN Health, October 29, 2018, https://www.cnn.com/2018/10/29/health/malaria-sniffing-dogs-study-africa /index.html. Megan Molteni, "The Science of the Sniff: Why Dogs Are Great Disease Detectors," *Wired*, October 30, 2018, https://www.wired.com/story/the-science-of-the-sniff-why-dogs-are -great-disease-detectors/.

13. Carina Wolff, "7 Unexpected Things Your Body Odor Is Trying to Tell You, According to Chinese

Medicine," *Bustle*, April 4, 2018, https://www.bustle.com/p/7-unexpected-things-your-body-odor-is-trying-to-tell-you-according-to-chinese-medicine-8649494.

14. Le Guérer, *Scent*, 23.

15. I sourced most of the facts in this section from Mark R. Smith, "Transcending, othering, detecting: Smell, premodernity, modernity." *Postmedieval: A Journal of Medieval Cultural Studies* 3 (2012), 380–90.

16. Kate Fox and Social Issues Research Centre (SIRC), "*The Smell Report: An Overview of Facts and Findings*," http://www.sirc.org/publik/smell_culture.html.

17. Hamidaddin, "Nose kiss, anyone?"

18. Shaunacy Ferro, "How Does Deodorant Work?" *Mental Floss*, September 24, 2015, https://www.mentalfloss.com/article/68960/how-does-deodorant-work.

19. Luke Dormiehl, "A new discovery could make next-gen deodorants way more effective," *Digital Trends*, July 23, 2018. https://www.digitaltrends.com/cool-tech/next-gen-deodorant-more-effective/.

20. Herz, *The Scent of Desire,* 18. See also Marta Zaraska, "The Sense of Smell in Humans is More Powerful Than We Think," *Discover Magazine*, October 11, 2017, and Tori Rodriguez, "Partners Can Smell Each Other's Emotions," *Scientific American*, January 1, 2012, https://www.scientificamerican.com/article/you-smell-angry/.

New Car

1. https://www.freep.com/story/money/cars/ford/2018/11/19/ford-new-car-smell-patent/2027822002/.

2. "They hate 'new car smell' in China," *Quartz*, January 27, 2018, video: https://www.youtube.com/watch?v=XgsZcnahVK8&feature=youtu.be&ab_channel=Quartz.

3. https://www.freep.com/story/money/cars/ford/2018/11/19/ford-new-car-smell-patent/2027822002/.

Cannabis

1. Benjamin Mueller, Robert Gebeloff, and Sahil Chinoy, "Surest Way to Face Marijuana Charges in New York: Be Black or Hispanic," *The New York Times*, May 13, 2018, https://www.nytimes.com/2018/05/13/nyregion/marijuana-arrests-nyc-race.html.

2. Jiachuan Wu and Daniella Silva, "MAP: See the states where marijuana is legal," *NBC News*, updated November 4, 2020, https://www.nbcnews.com/news/us-news/map-see-if-marijuana-legal-your-state-n938426. As of this writing, fifteen states plus D.C. and two territories have legalized recreational cannabis use, while thirty-four states and two territories allow medical usage.

3. Michael Rubinkam, "In era of legal pot, can police still search cars based on odor?" *PBS News Hour* (via Associated Press), September 13, 2019, https://www.pbs.org/newshour/nation/in-era-of-legal-pot-can-police-still-search-cars-based-on-odor.

4. "A Tale of Two Countries: Racially Targeted Arrests in the Era of Marijuana Reform," ACLU, April 17, 2020, https://www.aclu.org/news/criminal-law-reform/a-tale-of-two-countries-racially -targeted-arrests-in-the-era-of-marijuana-reform/.

5. Patrick Matthews, *Cannabis Culture: A Journey Through Disputed Territory* (United Kingdom: Bloomsbury, 2000), 2.

6. Dipak Hemraj, "Cannabis Sativa, Indica and Ruderalis," *Leafwell*, November 27, 2017, https ://leafwell.co/blog/cannabis-sativa-indica-and-ruderalis/. A third type of cannabis exists called ruderalis—the name derives from the Latin word for "rope." It has low concentrations of THC and CBD and thus few medicinal applications, but it's useful for cross-pollination and breeding.

7. Martin Booth, *Cannabis: A History* (New York: Picador, 2003), 7, 11. See also Avery N. Gilbert and Joseph A. DiVerdi, "Consumer perceptions of strain differences in Cannabis aroma," PLoS ONE 13, no. 2 (2018): e0192247, https://doi.org/10.1371/journal.pone.0192247.

8. Martin A. Lee, "What are Terpenes?," Project CBD, March 1, 2014, https://www.projectcbd.org /science/terpenes/terpenes-smell-mystery.

9. Aisha Hassan, "Quartz Obsession: Lavender," *Quartz*, November 7, 2018, https://qz.com/emails /quartz-obsession/1453721/. JoAnna Klein, "Lavender's Soothing Scent Could Be More Than Just Folk Medicine," The New York Times, October 23, 2018, https://www.nytimes.com/2018 /10/23/science/lavender-scent-anxiety.html?smid=tw-nytimes&smtyp=cu.

10. Seth Matlins and Eve Epstein, with illustrations by Ann Pickard, *The Scratch & Sniff Book of Weed* (New York: Abrams Image, 2017), 5–6, and Bailey Rahn, "What are cannabis terpenes and what do they do?" Leafly, February 12, 2014, updated October 1, 2019, https://www.leafly.com/news /cannabis-101/terpenes-the-flavors-of-cannabis-aromatherapy.

11. This entire mini-history comes courtesy of Booth, *Cannabis,* 2, 22–26, 48, 108, 114–19, 137–39, 154–57, 196–202, 207–209, 229–30.

12. Lisa Rough, "What is Bhang? A History Lesson and a Recipe," Leafly, March 21, 2017, https ://www.leafly.com/news/lifestyle/what-is-bhang-history-and-recipes.

Cash

1. Kelly Crow, "What Gives Money Its Distinctive Smell? One Chemist Tried to Find Out," *The Wall Street Journal*, updated January 18, 2017, https://www.wsj.com/articles/a-chemist-has-captured -the-worlds-most-elusive-fragrancethe-smell-of-dollar-bills-1484754108. Some countries— including Australia, the United Kingdom, and Canada—are switching some or all of their denom-inations to polymer-based "plastic" notes. Much more durable, these banknotes can be wiped clean with a damp cloth. Presumably these won't smell at all. See Neil Savage, "Here Comes the Plastic Money," *MIT Technology Review*, March 23, 2012, https://www.technologyreview .com/s/427308/here-comes-the-plastic-money/. CCL, maker of Guardian polymers for world currencies, explains their benefits in this product video: https://www.cclsecure.com/why -guardian/.

2. Matt Soniak, "Why Do Coins Make Your Hands Smell Funny?" *Mental Floss*, July 19, 2010, https://www.mentalfloss.com/article/25226/why-do-coins-make-your-hands-smell-funny.

Gasoline

1. "Benzene and Cancer Risk," American Cancer Society, Cancer.org, updated January 5, 2016, https://www.cancer.org/cancer/cancer-causes/benzene.html.
2. https://www.drugabuse.gov/publications/drugfacts/inhalants, and "Toxic Substances Portal—Gasoline, Automotive," Agency for Toxic Substances & Disease Registry, updated October 21, 2014.
3. Jason Fernando, "Sweet Crude," Investopedia.com, updated October 5, 2020, https://www.investopedia.com/terms/s/sweetcrude.asp.
4. Michael Frank, "Summer-Blend vs Winter-Blend Gasoline: What's the Difference?" *Popular Mechanics*, October 15, 2012, https://www.popularmechanics.com/cars/a3180/summer-blend-vs-winter-blend-gasoline-whats-the-difference-13747431/.

7. Sharp and Pungent

Musk

1. Groom, *The Perfume Handbook*, 151–52. Constance Classen, David Howes, and Anthony Synnott, eds., *Aroma: The Cultural History of Smell* (London and New York: Routledge, 1994), 71–72. Muchembled, *Smells,* 118.
2. Groom, *The Perfume Handbook*, 45, 54. See also Mallory Locklear, *Discover Magazine*, "5 Icky Animal Odors That Are Prized By Perfumers," October 12, 2014, https://www.discovermagazine.com/mind/5-icky-animal-odors-that-are-prized-by-perfumers.
3. Muchembled, *Smells,* 4.
4. Anya H. King, *Scent from the Garden of Paradise: Musk and the Medieval Islamic World* (Leiden and Boston: Brill, 2017), 333.
5. Le Guérer, *Scent*, 9.
6. Veronique Greenwood, "The Bacterial Surprise in This Bird's Smell," *The New York Times*, November 10, 2019, https://www.nytimes.com/2019/11/10/science/birds-smell-bacteria.html.
7. Natasha Frost, "The Regional Scent Dialects That Help Otters Tell Friends From Strangers," *Atlas Obscura*, December 13, 2017, https://www.atlasobscura.com/articles/otters-odor-dialects-dialects.
8. David Barrie, "Birds might follow their noses home," *Popular Science*, May 28, 2019, https://www.popsci.com/bird-navigation-smell/. This is an excerpt from David Barrie, *Super Navigators: Exploring the Wonders of How Animals Find Their Way* (New York: The Experiment, 2019), a book-length exploration of how animals navigate great distances.

9. Veronique Greenwood, "Elephants May Sniff Out Quantities With Their Noses," *The New York Times*, June 9, 2019, https://www.nytimes.com/2019/06/04/science/elephants-smell-quantity.html.

Oranges

1. John McPhee, *Oranges* (New York: Farrar, Straus and Giroux, 1983), 10.
2. McPhee, *Oranges*, 113.
3. Groom, *The Perfume Handbook*, 159. Helena Attlee, *The Land Where Lemons Grow: The Story of Italy and Its Citrus Fruit* (Woodstock, VT: The Countryman Press, 2015), 127–29.
4. Attlee, *The Land Where Lemons Grow*, 128. Groom, *The Perfume Handbook*, 192.
5. Groom, *The Perfume Handbook*, 28.
6. Newman, *Perfume,* 10.
7. Attlee, *The Land Where Lemons Grow*, 159–61. See also Sarah Bouasse, "Bergamot, the prized Calabrian scent," *Nez* 5 (Spring/Summer 2018), 54–59, and Reinarz, *Past Scents,* 76.
8. John Irving, "It's the real thing: Italy's relationship with chinotto has been a bittersweet affair . . . ," *Gourmet Traveler*, February 16, 2014, https://www.gourmettraveller.com.au/news/drinks-news/chinotto-6552.
9. Attlee, *The Land Where Lemons Grow*, 111–12.
10. Attlee, *The Land Where Lemons Grow*, 114 .
11. David Karp, "The Secrets Behind Many Chefs' Not-So-Secret Ingredient," *The New York Times*, December 3, 2003, https://www.nytimes.com/2003/12/03/dining/the-secrets-behind-many-chefs-not-so-secret-ingredient.html, and http://slowsoak.com/japanese-yuzu-bath-complete-guide/.

Lavender

1. Roberta Wilson, *The Essential Guide to Essential Oils: The Secret to Vibrant Health and Beauty* (New York: Avery, 2002), 87–88. Lavender also symbolized mistrust to Victorian England—see Gretchen Scoble and Ann Field, *The Meaning of Flowers: Myth, Language and Lore* (San Francisco: Chronicle Books, 1998), 56.
2. "Lavender," *Oxford English Dictionary (OED)*, accessed December 30, 2020, https://www-oed-com.proxy.uchicago.edu/view/Entry/106369.
3. Reinarz, *Past Scents,* 81–82. The Netflix documentary series *(Un)Well*, season 1, episode 1 "Essential Oils" (2020) explores what scientific evidence tells us about aromatherapy's medical effectiveness as well as the limits of its powers.
4. Steven Kurutz, "Why Does Everything Smell, So Peacefully, of Lavender?" *The New York Times*, September 14, 2019, https://www.nytimes.com/2019/09/14/style/lavender.html.
5. Koulivand, Peir Hossein, et al., "Lavender and the nervous system," *Evidence-based complementary and alternative medicine* (2013): 681304, doi:10.1155/2013/681304, https://www.ncbi.nlm.nih.gov/pmc/articles/PMC3612440/.

Skunk

1. Unless otherwise noted, the facts in this chapter come from these two articles: Mollie Bloudoff-Indelicato, "Why Skunks Evolved Their Smelly Spray," *National Geographic*, March 11, 2014, https://blog.nationalgeographic.org/2014/03/11/why-skunks-evolved-their-smelly-spray/, and Alicia Ault, "Ask Smithsonian: What Makes Skunk Spray Smell So Terrible?" *Smithsonian Magazine*, June 11, 2015, https://www.smithsonianmag.com/smithsonian-institution/ask-smithsonian-what-makes-skunk-spray-smell-so-terrible-180955553/.
2. You can watch this skunk dance here: "Spotted skunk handstand," *Weird Science, BBC Wildlife*, video: https://www.youtube.com/watch?v=WTQc-WEb5h8#aid=P-KPn-m0_AA&ab_channel=BBCStudios.
3. "The Mating Habits of Skunks," Wildlife Animal Control, http://wildlifeanimalcontrol.com/skunkmate.html.

Beer

1. Most of the chapter's points came from my email interviews with Mikey Manning. See also John Palmer, *How to Brew: Ingredients, Methods, Recipes and Equipment for Brewing Beer at Home* (Boulder, CO: Brewers Publications, 2006), http://howtobrew.com/book/section-1/a-crash-course-in-brewing/brew-day.
2. Palmer, *How to Brew*, http://howtobrew.com/book/section-1/yeast/what-is-it.
3. From email interviews with Mikey Manning. See also Palmer, *How to Brew*, http://howtobrew.com/book/section-1/a-crash-course-in-brewing/fermentation. This chapter also benefited by watching Brooklyn Brewery brewmaster Garrett Oliver in "Beer Expert Guesses Which Beer is More Expensive," *Price Points* season 1, episode 13, Epicurious.com, December 13, 2018, https://www.epicurious.com/video/watch/cheap-vs-expensive-beer-expert-guesses-which-beer-is-more-expensive.

Ditto Sheets

1. "Ditto," Dictionary.com, https://www.dictionary.com/browse/ditto?s=t.
2. Walter Benjamin, *The Work of Art in the Age of Its Technological Reproducibility, and Other Writings on Media*, ed. Michael W. Jennings, Brigid Doherty, and Thomas Y. Levin (Cambridge, MA: Belknap Press, 2008), 23.
3. Benjamin, *The Work of Art in the Age of Its Technological Reproducibility, 23.*
4. "Antique Copying Machines," OfficeMuseum.com, https://www.officemuseum.com/copy_machines.htm. "Dead Medium: Spirit Duplicators," DeadMedia.org, http://www.deadmedia.org/notes/40/408.html.
5. Eric Zorn, "That ditto high is harder and harder to duplicate," *Chicago Tribune*, January 16, 2007, https://blogs.chicagotribune.com/news_columnists_ezorn/2007/01/ditto_machines_.html.

8. Salty and Nutty

Ocean

1. Benjamin Wolfe, "Why Does The Sea Smell Like The Sea?" *Popular Science*, August 19, 2014, https://www.popsci.com/seasmells/.
2. Aftel, *Fragrant,* 188–90. See also https://www.aftelier.com/Onycha-Tincture-p/bot-eo-onycha .htm.
3. Olivier R.P. David, "Calone," *Nez* 5 (Spring/Summer 2018): 18–19.

Ambergris

1. Melville, Herman, *Moby-Dick; or, The Whale* (London: Constable & Co., 1922; Bartleby.com, 2013), https://www.bartleby.com/91/92.html. Accessed December 31, 2020.
2. Technically ambergris is not poop; it's a fatty, cholesterol-rich substance that originates in the whale's digestive tract. Sometimes it's incorrectly described as "whale vomit." I cannot improve upon an explanation I got from Christopher Kemp, author of *Floating Gold: A Natural (& Unnatural) History of Ambergris* (Chicago and London: The University of Chicago Press, 2012). From our email correspondence: "It's definitely not vomit and, although it's not quite poop, it makes the same journey as poop. It's poop-like. It's pathologically poop-like. I mean, fresh black ambergris really smells like poop and it has a lot on common with poop and that's worth noting."
3. Kemp, *Floating Gold*, 11–15. See also this *New York Times* article about how scientists are extracting sperm-whale DNA from ambergris to study it: Joshua Sokol, "New Origin Story for Gross Blobs That Wash Up on Beaches," *The New York Times*, February 4, 2020, https://www.nytimes.com /2020/02/04/science/ambergris-sperm-whales-dna.html?referringSource=articleShare.
4. Holly Dugan, *The Ephemeral History of Perfume: Scent and Sense in Early Modern England* (Baltimore: Johns Hopkins University Press, 2011), 130.
5. Kemp, *Floating Gold,* 24–25.
6. Potter, *The Rose,* 244.
7. Kemp, *Floating Gold,* 76.
8. Kemp, *Floating Gold,* 61–66.
9. Kemp, *Floating Gold,* 79.

Play-Doh

1. David Kindy, "The Accidental Invention of Play-Doh," *Smithsonian Magazine*, November 12, 2019, https://www.smithsonianmag.com/innovation/accidental-invention-play-doh-180973527/.
2. Patent #US3167440A, "Plastic modeling composition of a soft, pliable working consistency," https://patents.google.com/patent/US3167440A/en. Thanks to Peter DiCola, Professor of Law at Northwestern University, for providing this link and fact-checking my grasp of IP law.

3. Shelley Morgan, "Why Can't I Patent a Fragrance?," LegalBeagle.com, https://legalbeagle.com/6162020-cant-patent-fragrance-.html. See also Rachel Siegel, "Remember how Play-Doh smells? U.S. trademark officials get it," *The Washington Post*, May 24, 2018, https://www.washingtonpost.com/news/business/wp/2018/05/24/remember-how-play-doh-smells-u-s-trademark-officials-get-it/, https://www.ipwatchdog.com/2017/12/21/scent-trademarks-complexities/id=91071/. Burr, *The Perfect Scent,* explores this illogic to fascinating effect—see pages 126–38. As the scent critic for *T: The New York Times Style Magazine*, he's certainly enough of an industry insider to know whereof he writes.

4. Nick Greene, "The 10 Current Scent Trademarks Currently Recognized by the U.S. Patent Office," *Mental Floss*, October 13, 2015, https://www.mentalfloss.com/article/69760/10-scent-trademarks-currently-recognized-us-patent-office.

5. Jaburg Wilk Attorneys at Law, "Hasbro Just Successfully Trademarked the Smell of Play-Doh," *JDSupra.com*, May 25, 2018, https://www.jdsupra.com/legalnews/hasbro-just-successfully-trademarked-62406/.

Wet Wool

1. Aussie Sheep and Wool Products, http://www.aussiesheepandwool.com.au/webcontent5.htm.

2. Dominic Greene, "Wool, wheat and wet weather," review of *The Last Wolf: The Hidden Springs of Englishness*, by Robert Winder, *The Spectator* (UK), August 12, 2017, https://www.spectator.co.uk/article/wool-wheat-and-wet-weather.

Peanut Butter

1. Corey Whelan, "Why Am I Always Craving Peanut Butter?" *Healthline*, updated May 23, 2018, https://www.healthline.com/health/craving-peanut-butter.

2. "How a Peanut Butter Test May Detect Alzheimer's," Cleveland Clinic, December 15, 2020, https://health.clevelandclinic.org/peanut-butter-test-may-detect-alzheimers/. Note that subsequent studies have yet to replicate this effect in other patient populations.

3. Sharon Begley, "Fever checks are a flawed way to flag Covid-19 cases. Experts say smell tests might help," STAT, July 2, 2020, https://www.statnews.com/2020/07/02/smell-tests-temperature-checks-covid19/.

4. Glacier Media, "Sniff test: How peanut butter could help identify COVID-19 carriers," *St. Albert Today*, April 25, 2020, https://www.stalberttoday.ca/beyond-local/sniff-test-how-peanut-butter-could-help-identify-covid-19-carriers-2282914.

5. Lynne Peeples, "Smell Tests for COVID-19 Are Coming to a College Near You," *Daily Beast,* August 16, 2020, https://www.thedailybeast.com/smell-tests-for-covid-19-are-coming-to-a-college-near-you.

6. Victoria Groce, "Can Smelling Peanuts Cause an Allergic Reaction?" Verywell Health, November 17, 2019, https://www.verywellhealth.com/peanut-allergy-smell-1324378, and Luis Villazon,

"Can you be allergic to a smell?" *BBC ScienceFocus*, https://www.sciencefocus.com/the-human
-body/can-you-be-allergic-to-a-smell/.

7. Paul Rincon, "Oldest material on Earth discovered," *BBC News*, January 13, 2020, https://www
.bbc.com/news/science-environment-51099609.

9. Tingling and Fresh

Snow

1. Anne Helmenstine, "Can You Smell Snow?" Science Notes, November 13, 2019, (updated January 19, 2020), https://sciencenotes.org/can-you-smell-snow/.

2. "Acoustical Properties of Snow," ActforLibraries.org, http://www.actforlibraries.org/acoustic
-properties-of-snow/, and W. Maysenhölder et al., "Microstructure and Sound Absorption of Snow," *Cold Regions Science and Technology* 83–84 (December 2012), 3–12, EBSCOhost, doi:10
.1016/j.coldregions.2012.05.001.

3. Jennifer Frazer, "Wonderful Things: Don't Eat the Pink Snow," *Scientific American*, July 9, 2013, https://blogs.scientificamerican.com/artful-amoeba/wonderful-things-dont-eat-the-pink
-snow/, and Joyce Gellhorn, *Song of the Alpine: The Rocky Mountain Tundra Through the Seasons* (Boulder, CO: Big Earth Publishing, 2002), 48, https://books.google.com/books?id=0PpLg9
mBJ8AC&lpg=PA48&ots=c4fvxg8q7s&dq=watermelon%20snow%20laxative&pg=PA48#v
=onepage&q=watermelon%20snow%20laxative&f=false.

Ginger

1. Turner, *Spice,* 193.
2. Turner, *Spice*, 195.
3. Turner, *Spice*, 185.
4. Turner, *Spice*, 192.
5. Turner, *Spice*, 192, 199.
6. Turner, *Spice*, 192.
7. Turner, *Spice*, 189.

Rosemary

1. Gary Allen, *Herbs: A Global History* (London: Reaktion Books, 2012), 20.
2. Marina Heilmeyer, *Ancient Herbs* (Los Angeles: Getty Publications, 2007), 86.
3. Malcolm Moore, "Eau de BC: the oldest perfume in the world," *The Telegraph* (UK), March 21, 2007, https://www.telegraph.co.uk/news/worldnews/1546277/Eau-de-BC-the-oldest-perfume
-in-the-world.html.
4. Groom, *The Perfume Handbook*, 107. See also Newman, *Perfume,* 38.

5. Lucienne A. Roubin, "Fragrant Signals and Festive Spaces in Eurasia," in *The Smell Culture Reader*, ed. Jim Drobnick (New York: Berg, 2006), 130–31.

6. Dugan, *The Ephemeral History of Perfume*, 101.

7. Gretched Scoble and Ann Field, *The Meaning of Flowers: Myth, Language and Lore* (San Francisco: Chronicle Books, 1998), 21.

8. Mandy Kirkby, *A Victorian Flower Dictionary* (New York: Ballantine Books), 133–34.

Pine

1. Mary M. Dusenbury and the Spencer Museum of Art, *Flowers, Dragons & Pine Trees: Asian Textiles in the Spencer Museum of Art* (New York and Manchester: Hudson Hills, 2004), 248.

2. Wolfram Eberhard, *A Dictionary of Chinese Symbols: Hidden Symbols in Chinese Life and Thought*, trans. G. L. Campbell (New York and London: Routledge & Kegan Paul Ltd, 2015), 292.

3. Deanna Conners, "Why pine trees smell so good," EarthSky, December 22, 2016, https://earthsky .org/earth/why-conifer-christmas-trees-pine-spruce-fir-smell-terpenes.

4. Alexandra Sifferlin, "Why Spring Is the Perfect Time to Take Your Workout Outdoors," *Time*, March 30, 2017, https://time.com/4718318/spring-exercise-workout-outside/.

5. Don J. Durzan, "Arginine, scurvy and Cartier's 'tree of life,'" *Journal of ethnobiology and ethnomedicine* 5, no. 5 (February 2, 2009), doi:10.1186/1746-4269-5-5, https://www.ncbi.nlm.nih.gov /pmc/articles/PMC2647905/.

10. Otherworldly

New Baby

1. Douglas Quenqua, "Ah, There's Nothing Like New Baby Smell," *The New York Times*, October 2, 2013, https://well.blogs.nytimes.com/2013/10/02/ah-theres-nothing-like-new-baby-smell/?mtrref=und efined&gwh=185E0424B4A3663C8553E219D9506E8A&gwt=pay&assetType=REGIWALL.

2. H. Varendi and R H Porter, "Breast odour as the only maternal stimulus elicits crawling towards the odour source," *Acta Paediatrica* 90, no. 4 (2001): 372–75, https://pubmed.ncbi.nlm.nih.gov /11332925.

3. M. Kaitz et al, "Mothers' recognition of their newborns by olfactory cues." *Developmental Psychobiology* 20, no. 6 (1987): 587–91, doi:10.1002/dev.420200604, https://pubmed.ncbi.nlm.nih .gov/3691966/.

4. Quenqua, "New Baby Smell."

5. Frontiers, "Mother nose best: Child body odor provides olfactory clues to developmental stages," *Medical Xpress*, March 4, 2020, https://medicalxpress.com/news/2020-03-mother-nose-child -body-odor.html.

6. Laurie L. Dove, "What causes 'old person' smell?" *How Stuff Works,* https://health.howstuffworks .com/mental-health/human-nature/perception/old-person-smell.htm.

Extinct Flowers

1. See "Resurrecting the Sublime" exhibit's About page, https://www.resurrectingthesublime.com/about.

2. Rowan Jacobsen, "Fragrant Genes of Extinct Flowers Have Been Brought Back to Life," *Scientific American*, February 2019, https://www.scientificamerican.com/article/fragrant-genes-of-extinct-flowers-have-been-brought-back-to-life/.

3. Alexandra Daisy Ginsberg emphasized that, for all the scientific rigor in this process, the result is still necessarily inconclusive. As she noted in an email to me, "We cannot be sure that this is the real smell: we don't know if the smell molecules were produced by the flowers, or present elsewhere in the plant, in what quantities, or whether the genes were switched on even! We cannot really know how the flowers smelled. Hibiscus as a genus don't really smell, as they are bird pollinated not insect pollinated. DNA can only tell us one story, not the whole story." Sissel Tolaas elaborates in the same email chain: "The core molecules are very 'characteristic & smelly' so most likely that can determine the SmellID of the plants. Of course the exact combination of the remaining molecules is hard to know precisely. This is also why I decided to display the smell molecules in various ways and combinations. BUT always the core ones are recognizable."

4. Carol Ann McCormick, "The Heartbreak of Psoralea," *North Carolina Botanic Garden (NCBG) Newsletter*, September-October 2007, http://www.herbarium.unc.edu/9-10-07.pdf.

An Invented Smell

1. More about Molecule 01 specifically: https://us.escentric.com/collections/escentric-molecules-01. Technically speaking, Molecule 01 consists of 100 percent Iso Gamma, a more concentrated variation of the original Iso E Super molecule. Iso E Super is a highly complex molecule which has undergone many, many isomeric transformations since its invention to isolate and concentrate its smell qualities. See Nicolas Armanino et al., "What's Hot, What's Not: The Trends of the Past 20 Years in the Chemistry of Odorants," *Angewandte Chemie* 59, no. 38 (September 14, 2020): 16310–44, https://doi.org/10.1002/anie.202005719, as well as Cornelius Nussbaumer, Georg Fráter, and Philip Kraft, "(±)-1-[(1R*,2R*,8aS*)-1,2,3,5,6,7,8,8a-Octahydro-1,2,8,8-tetramethylnaphthalen-2-yl]ethan-1-one: Isolation and Stereoselective Synthesis of a Powerful Minor Constituent of the Perfumery Synthetic Iso E Super®," *Helvetica Chimica Acta* 82, no. 7 (July 7, 1999), 1016–24, https://doi.org/10.1002/(SICI)1522-2675(19990707)82:7<1016::AID-HLCA1016>3.0.CO;2-Y.

2. Unless otherwise noted, all the facts from this chapter stem from two articles: Bettina Weber, "Erfindung der Düfte" ["Invention of Smells"], *Bolero,* October 2003, 70–71 (in German), and Matvey Yudov, "The History of Iso E Super in Perfumery," *Fragrantica*, https://www.fragrantica.com/news/The-History-of-Iso-E-Super-in-Perfumery-7729.html#:~:text=Among%20other%20synthetic%20perfumery%20ingredients,Super%20stands%20out%20a%20little.&text

=The%20history%20of%20Iso%20E%20Super%20began%20in%20the%201960s,In%20
1973%2C%20John%20B.

3. Williams, "Human nose can detect a trillion smells," https://www.sciencemag.org/news/2014/03
/human-nose-can-detect-trillion-smells, as well as Günther Ohloff, Wilhelm Pickenhagen, and
Philip Kraft, *Scent and Chemistry: The Molecular World of Odors* (Germany: Verlag Helvetica Chimica Acta and Wiley-VCH, 2012), 27.

4. Cengage, "Cinnamaldehyde," Encyclopedia.com, https://www.encyclopedia.com/science/academic
-and-educational-journals/cinnamaldehyde.

5. Newman, *Perfume,* 82. See also Chandler Burr, *The Perfect Scent: A Year Inside the Perfume Industry
in Paris and New York* (New York: Henry Holt and Company, 2007), *The Perfect Scent,* 117–26.

6. Newman, *Perfume,* 85–95, 100. Gilbert, *What the Nose Knows,* 26–28, 37.

7. The best deep-dive read about the modern business and art of creating perfumes is Chandler
Burr's *The Perfect Scent.* This chapter owes a lot to reading his work.

Ectoplasm

1. Marina Warner, "Ethereal Body: The Quest for Ectoplasm," *Cabinet Magazine*, Fall/Winter 2003,
http://www.cabinetmagazine.org/issues/12/warner.php. Most of this chapter's facts are sourced
from this excellent article.

2. https://encyclopedia2.thefreedictionary.com/Ectoplasma.

3. Daniel Engber, "What's the Deal with Paranormal Ectoplasm?" *Popular Science*, August 5, 2015,
https://www.popsci.com/whats-deal-paranormal-ectoplasm/.

4. "Ectoplasm," The Free Dictionary, https://encyclopedia2.thefreedictionary.com/Ectoplasma. See
also Kate Kershner, "What is ectoplasm?" *How Stuff Works*, April 7, 2015, https://science
.howstuffworks.com/science-vs-myth/unexplained-phenomena/ectoplasm.htm.

The Odor of Sanctity

1. Ben Gazur, "The Smell of Saintliness," Wellcome Collection Stories, November 12, 2019, https
://wellcomecollection.org/articles/XamCsxAAACAAqWIm.

2. "Module #103: Polycarp's Martyrdom," Christian History Institute's Early Church online studies
series, https://christianhistoryinstitute.org/study/module/polycarp.

3. Constance Classen, "The Breath of God: Sacred Histories of Scent," in *The Smell Culture Reader*,
ed. Jim Drobnick (New York: Berg, 2006), 376–79. See also Le Guérer, *Scent*, 122–23.

4. "St. Simeon Stylites," *Encyclopedia Britannica*, https://www.britannica.com/biography/Saint
-Simeon-Stylites.

5. Stephen Beale, "What does sanctity smell like?" *Catholic Exchange*, August 11, 2014, https
://catholicexchange.com/sanctity-smell-like.

6. Suzanne Evans, "The Scent of a Martyr," *Numen* 49, no. 2 (2002): 201–202. She is quoting a story
of St. Lydwine penned by the nineteenth-century French Decadent author Joris-Karl Huysmans.

7. "Al-Ḥusayn ibn ʿAlī," *Encyclopedia Britannica*, https://www.britannica.com/biography/al-Husayn
-ibn-Ali-Muslim-leader-and-martyr.

8. Evans, "The Scent of a Martyr," 204–207.

9. Andrea Tapparo et al., "Chemical characterisation, plant remain analysis and radiocarbon dating
of the Venetian 'Manna di San Nicola,'" *Annali di chimica* 92, no. 3 (2002): 327–32. https://pubmed
.ncbi.nlm.nih.gov/12025516/.

Old Books

1. American Chemical Society, "That Old Book Tells Its Secrets," Technology Networks, November
19, 2019, https://www.technologynetworks.com/analysis/news/that-old-book-smell-tells-its
-secrets-327385, and "What Causes the Smell of New & Old Books?" Compound Interest, June
1, 2014, https://www.compoundchem.com/2014/06/01/newoldbooksmell/. See also Mark Kur-
lansky, *Paper: Paging Through History* (United States: W. W. Norton, 2016), 250–55.

2. Erin Blakemore, "The Quest to Better Describe the Scent of Old Books," *Smithsonian Magazine*,
April 7, 2017, https://www.smithsonianmag.com/smart-news/the-quest-better-describe-scent
-old-books-180962819/. Veronique Greenwood, "Can an Archive Capture the Scents of an Entire
Era?" *The Atlantic*, May 15, 2017, https://www.theatlantic.com/science/archive/2017/05/smell
-archive/526575/. Daniel A. Gross, "How Smell Tests Can Help Museums Conserve Art and Artifacts," *Hyperallergic*, March 12, 2018, https://hyperallergic.com/431947/smell-tests-can-help
-museums-conserve-art-artifacts/.

Conclusion

1. Franklin Mariño-Sánchez et al., "Smell training increases cognitive smell skills of wine tasters
compared to the general healthy population. The WINECAT Study," *Rhinology* 48, no. 3 (September 2010): 273–76, doi: 10.4193/Rhin09.206, https://www.researchgate.net/publication
/47636697_Smell_training_increases_cognitive_smell_skills_of_wine_tasters_compared_to
_the_general_healthy_population_The_WINECAT_Study.

2. Asifa Majid, "Humans Are Neglecting Our Sense of Smell. Here's What We Could Gain By Fixing
That," *Time*, March 7, 2018, https://time.com/5130634/sense-smell-milk-expiration-industrial
-revolution/.

SELECTED BIBLIOGRAPHY

CULTURAL HISTORIES

Ackerman, Diane. *A Natural History of the Senses.* New York: Vintage Books, 1990.

Barwich, A. S. *Smellosophy: What the Nose Tells the Mind.* Cambridge: Harvard University Press, 2020.

Classen, Constance, David Howes, and Anthony Synnott, eds. *Aroma: The Cultural History of Smell.* London and New York: Routledge, 1994.

Classen, Constance. *Worlds of Sense: Exploring the Senses in History and Across Cultures.* London and New York: Routledge, 1993.

Corbin, Alain. *The Foul and the Fragrant: Odor and the French Social Imagination.* Cambridge: Harvard University Press, 1986.

Drobnick, Jim, ed. *The Smell Culture Reader.* New York: Berg, 2006.

Le Guérer, Annick. *Scent: The Essential and Mysterious Powers of Smell.* Translated by Richard Miller. New York: Kodansha America, 1994.

Muchembled, Robert. *Smells: A Cultural History of Odours in Early Modern Times.* Translated by Susan Pickford. Cambridge, UK, and Medford, MA: Polity, 2020.

Reinarz, Jonathan. *Past Scents: Historical Perspectives on Smell.* Urbana, Chicago, and Springfield: University of Illinois Press, 2014.

SCIENCE OF OLFACTION

Burr, Chandler. *The Emperor of Scent: A Story of Perfume, Obsession and the Last Mystery of the Senses.* New York: Random House Publishing Group, 2003.

Cobb, Matthew. *Smell: A Very Short Introduction.* Oxford: Oxford University Press, 2020.

Gilbert, Avery. *What the Nose Knows: The Science of Scent in Everyday Life.* N.P.: Synesthetics, Inc., 2014.

Herz, Rachel. *The Scent of Desire: Discovering Our Enigmatic Sense of Smell.* New York: Harper Perennial, 2007.

Horowitz, Alexandra. *Being a Dog: Following the Dog Into a World of Smell.* New York: Scribner, 2016.

McGee, Harold. *Nose Dive: A Field Guide to the World's Smells.* New York: Penguin Press, 2020.

Pelosi, Paolo. *On the Scent: A Journey Through the Science of Smell.* Oxford: Oxford University Press, 2016.

Stoddart, Michael. *Adam's Nose, and the Making of Humankind.* London: Imperial College Press, 2015.

Turin, Luca. *The Secret of Scent: Adventures in Perfume and the Science of Smell.* New York: Harper Perennial, 2007.

Vroon, Piet, with Anton van Amerangen and Hans De Vries. *Smell: The Secret Seducer.* Translated by Paul Vincent. New York: Farrar, Straus and Giroux, 1997.

PERFUMES

Aftel, Mandy. *Essence & Alchemy: A Book of Perfume.* Layton, UT: Gibbs Smith, 2008.

———. *Fragrant: The Secret Life of Scent.* New York: Riverhead Books, 2014.

Burr, Chandler. *The Perfect Scent: A Year Inside the Perfume Industry in Paris and New York.* New York: Henry Holt and Company, 2007.

Dugan, Holly. *Scent and Sense in Early Modern England.* Baltimore: Johns Hopkins University Press, 2011.

Klanten, Robert, Carla Seipp, and Santiago Rodriguez Tarditti, eds. *The Essence: Discovering the World of Scent, Perfume & Fragrance.* Berlin: Die Gestalten Verlag, 2020.

Groom, Nigel. *The Perfume Handbook.* London: Chapman & Hall, 1992.

Newman, Cathy, with photography by Robb Kendrick. *Perfume: The Art and Science of Scent.* Washington, DC: National Geographic Society, 1998.

Stewart, Susan. *Cosmetics & Perfumes in the Roman World.* Gloucestershire, UK: Tempus Publishing, 2007.

Turin, Luca, and Tania Sanchez. *Perfumes: The Guide.* New York: Viking, 2008.

FLOWERS, HERBS, AND SPICES

Allen, Gary. *Herbs: A Global History.* London: Reaktion Books, 2012.

Czarra, Fred. *Spices: A Global History.* London: Reaktion Books, 2009.

Heilmeyer, Marina. *Ancient Herbs.* Los Angeles: Getty Publications, 2007.

Kirkby, Mandy. *A Victorian Flower Dictionary.* New York: Ballantine Books, 2011.

Scoble, Gretchen, and Ann Field. *The Meaning of Flowers: Myth, Language and Lore.* San Francisco: Chronicle Books, 1998.

Turner, Jack. *Spice: The History of a Temptation.* New York: Knopf, 2004.

Wells, Diana, with illustrations by Ippy Patterson. *100 Flowers & How They Got Their Names.* Chapel Hill, NC: Algonquin Books of Chapel Hill, 1997.

MEMOIRS

Birnbaum, Molly. *Season to Taste: How I Lost My Sense of Smell and Found My Way.* New York: Ecco, 2011.

Blodgett, Bonnie. *Remembering Smell: A Memoir of Losing—and Discovering—The Primal Sense.* Boston: Houghton Mifflin Harcourt, 2010.

Ellena, Jean-Claude. *The Diary of a Nose: The Year in the Life of a Parfumeur.* Translated by Adriana Hunter. New York: Rizzoli Ex Libris, 2013.

Harad, Alyssa. *Coming to My Senses: A Story of Perfume, Pleasure, and an Unlikely Bride.* New York: Viking, 2012.

AROMATHERAPY

Farrer-Halls, Gill. *The Aromatherapy Bible: The Definitive Guide to Using Essential Oils.* New York: Sterling Publishing Co., Inc., 2005.

Wilson, Roberta. *The Essential Guide to Essential Oils: The Secret to Vibrant Health and Beauty.* New York: Avery, 2002.

BOOKS ON INDIVIDUAL SMELLS

Ambergris

Kemp, Christopher. *Floating Gold: A Natural (& Unnatural) History of Ambergris.* Chicago and London: The University of Chicago Press, 2012.

Beer

Palmer, John. *How to Brew: Ingredients, Methods, Recipes and Equipment for Brewing Beer at Home.* Boulder, CO: Brewers Publications, 2006.

Nachel, Marty. *Beer for Dummies.* Hoboken, NJ: Wiley, 2012.

Camphor

Donkin, R. A. *Dragon's Brain Perfume: An Historical Geography of Camphor.* Leiden and Boston: Brill, 1999.

Cannabis

Abel, Ernest L. *Marihuana: The First Twelve Thousand Years.* New York: Plenum Press, 1980.

Booth, Martin. *Cannabis: A History.* New York: Picador, 2003.

Matlins, Seth, and Eve Epstein, with illustrations by Ann Pickard. *The Scratch & Sniff Book of Weed.* New York: Abrams Image, 2017.

Cannon Fire

Smith, Mark M. *The Smell of Battle, The Taste of Siege: A Sensory History of the Civil War.* Oxford: Oxford University Press, 2015.

Cheese

Dalby, Andrew. *Cheese: A Global History.* London: Reaktion Books, 2009.

LeMay, Eric. *Immortal Milk: Adventures in Cheese.* New York: Free Press, 2010.

Percival, Bronwen, and Francis Percival. *Reinventing the Wheel: Milk, Microbes and the Fight for Real Cheese.* Oakland: University of California Press, 2017.

The Cheese Nun: Sister Noella's Voyage of Discovery. DVD. Directed by Pat Thompson. Alexandria, VA: PBS Home Video, 2006.

Chocolate

Moss, Sarah, and Alexander Badenoch. *Chocolate: A Global History.* London: Reaktion Books, 2009.

Frankincense and Myrrh

Groom, Nigel. *Frankincense and Myrrh: A Study of the Arabian Spice Trade.* Harlow, Essex, UK, and Beirut: Longman Group and Librairie du Liban, 1981.

Jasmine

Grasse, Marie-Christine. *Jasmine: Flower of Grasse.* Parkstone Publishers & Musée Internationale de la Perfume, 1996.

Lavender

Ralston, Jeannie. *The Unlikely Lavender Queen: A Memoir of Unexpected Blossoming.* New York: Broadway Books, 2008.

Musk

King, Anya H. *Scent from the Garden of Paradise: Musk and the Medieval Islamic World.* Leiden and Boston: Brill, 2017.

The Odor of Sanctity

Harvey, Susan Ashbrook. *Scenting Salvation: Ancient Christianity and the Olfactory Imagination.* Berkeley: University of California Press, 2006.

Oranges

Attlee, Helena. *The Land Where Lemons Grow: The Story of Italy and Its Citrus Fruit.* Woodstock, VT: The Countryman Press, 2015.

McPhee, John. *Oranges.* New York: Farrar, Straus and Giroux, 1983.

Oud

Morita, Kiyoko. *The Book of Incense: Enjoying the Traditional Art of Japanese Scents.* Tokyo and New York: Kodansha International, 1992.

Al-Woozain, Mohamed. *Scent of Heaven: On the Trail of Oud.* Al Jazeera documentary, 2016. https://interactive.aljazeera.com/aje/2016/oud-agarwood-scent-from-heaven/index.html.

Pencils

Petroski, Henry. *The Pencil: A History of Design and Circumstance.* London and Boston: Faber and Faber, 2003.

Weaver, Caroline. *The Pencil Perfect: The Untold Story of a Cultural Icon.* Berlin: Gestalten, 2017.

Roses

Bernhardt, Peter. *The Rose's Kiss: A Natural History of Flowers.* Washington, DC, and Covelo, CA: Island Press and Shearwater Books, 1999.

Potter, Jennifer. *The Rose: A True History.* London: Atlantic Books, 2010.

Tea

Saberi, Helen. *Tea: A Global History.* London: Reaktion Books, 2010.

Tobacco

Gately, Iain. *Tobacco: A Cultural History of How an Exotic Plant Seduced Civilization.* New York: Grove Press, 2001.

Goodman, Jordan, ed. *Tobacco in History and Culture: An Encyclopedia.* Detroit: Thomson Gale, 2005.

Truffles

Hall, Ian R. *Taming the Truffle: The History, Lore and Science of the Ultimate Mushroom.* Portland, OR: Timber Press, 2007.

Jacobs, Ryan. *The Truffle Underground: A Tale of Mystery, Mayhem and Manipulation in the Shadowy Market of the World's Most Expensive Fungus.* New York: Clarkson Potter, 2019.

Maser, Chris. *Trees, Truffles and Beasts: How Forests Function.* New Brunswick, NJ: Rutgers University Press, 2008.

Wells, Patricia. *Simply Truffles: Recipes and Stories That Capture the Essence of the Black Diamond.* New York: William Morrow, 2011.

Vanilla

Rain, Patricia. *Vanilla: The Cultural History of the World's Favorite Flavor and Fragrance.* New York: Jeremy P. Tarcher, 2004.

Ecott, Tim. *Vanilla: Travels in Search of the Ice Cream Orchid.* New York: Grove Press, 2004.

Wine

Betts, Richard, with Crystal English Sacca, illustrated by Wendy MacNaughton. *The Essential Scratch & Sniff Guide to Becoming a Wine Expert.* New York: Rux Martin, 2013.

Bosker, Bianca. *Cork Dork: A Wine-Fuelled Adventure Among the Obsessive Sommeliers, Big Bottle Hunters and Rogue Scientists Who Taught Me to Live for Taste.* New York: Penguin Books, 2017.

Chartier, François. *Taste Buds and Molecules: The Art and Science of Food, Wine and Flavor.* Translated by Levi Reiss. Hoboken, NJ: John Wiley & Sons, 2012.

Robinson, Jancis. *How to Taste: A Guide to Enjoying Wine.* New York: Simon & Schuster, 2008.

Shepherd, Gordon M. *Neurogastronomy: How the Brain Creates Flavor and Why It Matters.* New York: Columbia University Press, 2012.